Creating a STEM Culture
for Teaching and Learning

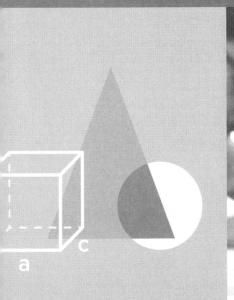

Creating a STEM Culture

for Teaching and Learning

Jeff Weld

National Science Teachers Association

Arlington, Virginia

National Science Teachers Association

Claire Reinburg, Director
Rachel Ledbetter, Managing Editor
Amanda Van Beuren, Associate Editor
Donna Yudkin, Book Acquisitions Manager

ART AND DESIGN
Will Thomas Jr., Director, Cover and
Interior Design

PRINTING AND PRODUCTION
Catherine Lorrain, Director

NATIONAL SCIENCE TEACHERS ASSOCIATION
David L. Evans, Executive Director
David Beacom, Publisher

1840 Wilson Blvd., Arlington, VA 22201
www.nsta.org/store
For customer service inquiries, please call 800-277-5300.

NSTA is committed to publishing material that promotes the best in inquiry-based science education. However, conditions of actual use may vary, and the safety procedures and practices described in this book are intended to serve only as a guide. Additional precautionary measures may be required. NSTA and the author do not warrant or represent that the procedures and practices in this book meet any safety code or standard of federal, state, or local regulations. NSTA and the author disclaim any liability for personal injury or damage to property arising out of or relating to the use of this book, including any of the recommendations, instructions, or materials contained therein.

PERMISSIONS

Library of Congress Cataloging-in-Publication Data
Names: Weld, Jeff, 1960- author.
Title: Creating a STEM culture for teaching and learning / Jeff Weld.
Description: Arlington, VA : National Science Teachers Association, [2017] | Includes bibliographical references.
Identifiers: LCCN 2017006107 (print) | LCCN 2017025140 (ebook) | ISBN 9781681403977 (e-book) | ISBN 9781681403960 (print)
Subjects: LCSH: Science--Study and teaching. | Technology--Study and teaching. | Engineering--Study and teaching. | Mathematics--Study and teaching.
Classification: LCC Q181 (ebook) | LCC Q181 .W4366 2017 (print) | DDC 507.1--dc23
LC record available at *https://lccn.loc.gov/2017006107*

Contents

Contents

Contents

About This Book

*C*reating a STEM Culture for Teaching and Learning* is small part histor-
ical treatise and most part contemporary profile. Its content reaches
back 30 years when SMET and STS were the operative abbrevia-
tions for a transformative, nascent vision for education. And it stretches
forward decades, painting the horizon that awaits this generation of learners. Between those outward points on the time line of the most exciting
era in science and mathematics education, the story of STEM unfolds
as a *how-to, can do, who's who, you too* manual and memoir based on the
experiences of leaders who walk the talk. Each chapter is sprinkled with
light-hearted illustrative case studies and metaphors along with real-time
exemplary vignettes to complement the topic at hand. What this book is
not is a compendium of all known programs across the country, although it
highlights a significant few. It is also *not* a research tome that reads like a
doctoral dissertation, although a number of studies and references are cited
in support of claims.

At its essence the book channels the wisdom and practice of hundreds
of professionals in education, business, nonprofit, and state and local government who together have built nationally recognized models for STEM
from the level of theory and policy to street-level credibility of the classroom. It is for teachers, administrators, business partners, community
members, parents, scholars, and policy makers who seek to be up-to-speed
on the many elements of STEM about which this book was compiled.
Inspired and prepared students will be the measure of its success.

Acknowledgments

A decade of personal encounters, life experiences, research studies, and probing observations amalgamate in *Creating a STEM Culture for Teaching and Learning*. Its existence owes to trusted mentors who said such a book is needed and that I am the one to do it. It traces back to visionary school leaders who tolerated my STS teaching forays in the pre-STEM science classroom and to university deans and department heads who later embraced the maverick-to-mainstream STEM transformation of courses and programs to which I was entrusted. It is a prominent credit to leaders in Iowa government, most notably she who has the last word in Chapter 9. Governor Kim Reynolds positioned me to direct, alongside amazing teammates, one of the most comprehensive, dynamic, and productive statewide STEM projects from which the purview between these covers draws. This book is a stew of thoughts and actions of 17 direct contributors and countless more who indirectly shaped education's STEM awakening. Should any of those mentors and thought leaders wonder to whom I refer, I look forward to dispelling their wonder.

Bravo to NSTA Press for advancing STEM education by welcoming the concept of this book and for nurturing its refinement toward public consumption.

Finally, the investment of hundreds of hours over countless nights and weekends in producing this book asked the greatest contribution of those closest to me. My sons David (who provided early, key reviews) and Andrew (with his new bride Madison) adapted gracefully to my schedule curveballs, all the while inspiring me by their devotion. But it is my life partner Mary who attended to all things not STEM to keep us happy, healthy, and connected during my single-mindedness. She also deserves credit for many of the bright ideas attributed to me over the course of Iowa's STEM journey. Her prints are all over this work.

About the Author

Jeff Weld, PhD, has directed the acclaimed Iowa Governor's STEM Advisory Council since its inception in 2011. On extended leave from a faculty position in the Department of Biology at the University of Northern Iowa where he specializes in science education, Weld is a frequent speaker, consultant, and writer on the national STEM scene. Formerly an award-winning high school science teacher in Mission, Texas; Kirkwood, Missouri; and Pella, Iowa; he draws inspiration for this book from the hundreds of colleagues and friends navigating the uncharted but promising waters of STEM education together. Many helped him account for the essential elements of STEM that follow.

CHAPTER 1

The STEM Imperative

There is a moral and economic underpinning to the STEM imperative. Anything short of the cultivation of a culture for STEM is insufficient. Success requires a systemic rather than a piecemeal approach to STEM education. All dimensions of schooling must be on the table—teacher preparation, scheduling, school–parent relations, professional development, curriculum, assessment, the disciplines, physical space, administrator support, business and community engagement, and of course budgets. STEM is ushering an invigorating evolution of education.

Rise of STEM

STEM is to education what Starbucks is to coffee—a game-changing, future-focused refinement on a basic good thing. Why is STEM taking the world by storm? Some 47 U.S. states have recently launched Science-Technology-Engineering-Mathematics initiatives according to STEMconnector, as have hundreds of localities, businesses, and government agencies. Consider the federal government sectors alone: the Army, the Navy, the Patent and Trademark Office, the NOAA, the Department of Labor, the Department of Transportation, as well as the Department of Education and others, each have their own STEM strategy and

programs. Across the globe STEM has become a continental priority, emerging as an educational imperative in Bahrain, Algeria, Taiwan, Australia, Korea, South Africa, and most nations in between. Fueling the frenzy is a steady parade of blue ribbon commissions and think-tank reports from organizations including the U.S. Chamber of Commerce, the National Governors' Association, the Carnegie Corporation, the National Research Council, the President's Council of Advisors on Science and Technology, and more, each alerting their readers to the promise of STEM and the perils of failing to act. STEM is big. The report of the President's Council of Advisors on Science and Technology (PCAST) states,

> *The success of the United States in the 21st century—its wealth and welfare—will depend on the ideas and skills of its population. These have always been the Nation's most important assets. As the world becomes increasingly technological, the value of these national assets will be determined in no small measure by the effectiveness of science, technology, engineering, and mathematics (STEM) education in the United States (2010).*

STEM is doing for mainstream math and science what school buses did for leveling educational opportunity in the 1920s—creating access to gateway courses for more students, and making such courses matter again. In communities, STEM is a nonpartisan, public-private sector unifier because there is something for every stakeholder group to like—teachers, business leaders, parents, policy makers, and most of all, students. Teachers are empowered by STEM to integrate concepts across disciplines and to put kids in the driver's seat as active learners. The business community gains access to teachers and classrooms as partners helping to shape meaningful lessons applied to the workplace. Parents enjoy dinner time debriefs about what happened at school today when their children

link lessons to cool jobs in town. Policy makers embrace broader constituencies for school funding when workforce and economic development advocates see education as an essential contributor. And students see school as a launch pad to life rather than an exercise in endurance.

Underscoring the mounting STEM priority for the nation, the 2015 federal Every Student Succeeds Act (ESSA) explicitly mandates attention to and support of STEM teaching professional development and student learning opportunities in Title II and Title IV of the Act:

> *STEM Master Teacher Corps (ESEA, as amended by ESSA, title II, section 2245): From funds reserved for title II national activities, grants may be awarded to: (1) SEAs to enable them to support the development of a statewide STEM master teacher corps or (2) SEAs or nonprofit organizations in partnership with SEAs to support the implementation, replication or expansion of effective STEM professional development programs in schools across the State through collaboration with school administrators, principals, and STEM educators ...*

> *Activities to Support Well-Rounded Educational Opportunities (ESEA, as amended by ESSA, title IV, section 4107): Each LEA or consortium of LEAs that receives an allocation under section 4105(a) shall use a portion of such funds to develop and implement programs and activities that support access to a well-rounded education and may include programming and activities that improve instruction and student engagement in science, technology, engineering, and mathematics, including computer science, such as:*

> a. *Increasing access for students through grade 12 who are members of groups underrepresented in STEM fields;*

> b. *Supporting the participation of low-income students in nonprofit competitions related to STEM subjects;*

> c. *Providing hands-on learning and exposure to STEM subjects and supporting the use of field-based or service learning to enhance students' understanding of the STEM subjects;*

> d. *Supporting the creation and enhancement of STEM-focused specialty schools;*

> e. *Facilitating collaboration among school, after-school program, and informal program personnel to improve the integration of programming and instruction in STEM subjects; and*

> f. *Integrating other academic subjects including the arts, into STEM subject programs to increase participation in STEM subjects, improve attainment of skills related to STEM, and promote well-rounded education. ...*

21st Century Community Learning Centers (ESEA, as amended by ESSA, title IV, section 4205): Each eligible entity that receives an award under section 4204 may use the award to carry out a broad array of expanded learning program activities that advance student academic achievement and support student success, including programs that build skills in STEM, including computer science, and that foster innovation in learning by supporting nontraditional STEM education teaching methods. (Every Student Succeeds Act, Pub. L. No. 114-95 § 114 Stat. 1177 [2015–2016])

To date, those STEM provisions of the Every Student Succeeds Act remain in place under the succeeding administration, though Congress recently voted to relax or suspend some of the rules and regulations associated with monitoring and enforcing the Act (see Goldstein 2017).

STEM as a Stress Response

To be part of America's STEM movement, perhaps more appropriately an *awakening*, is to dwell in a very rare confluence of three streams of expectancy:

1. Teaching aligned to best practice

2. A workforce sector mandate for more useful graduates

3. A societal thirst for citizens capable of grappling with the big challenges of our times, many with roots in STEM

If these expectancies on the part of stakeholders—parents, employers, policy makers, and in fact educators themselves—were being met during the past decade or two, it would be difficult for a disruptive educational innovation to make in-roads. But of course STEM is making vigorous headway. It is a stress response in the same way that a job change or a new hobby or exercise routine answers an internal nagging stressor to change things up in our personal lives. The education system at large, which includes not only schools and teachers but also their clients who inherit the graduates, namely higher education, employers, and communities, is under stress. Today's college freshmen too often require remediation, according to ACT's *Condition of College and Career Readiness 2015* (ACT 2015). Today's entry-level employees lack basic work skills, according to a report from the Society for Human Resource Management, *Are They Really Ready to Work?* (SHRM 2006). And, the millennial generation is said to lack functional civic capacities that would equip them to make wise decisions at the voting booth, the jury stand, or the grocery store, according to a cover story in *Time* called "Millennials: The Me Me Me Generation" (Stein 2013).

As hard as everyone in education is working, the product is believed not to be living up to expectations. And in the realm of public sentiment on education, perceptions become reality. STEM has arisen as the game-changing response to a rapidly spreading belief that schools must step up the production of college and career-ready graduates able and eager to thrive in rapidly evolving times. STEM is that New Year's Eve resolution that things are going to be different, better. And evidence has accumulated to make the case that it delivers. But unlike so many resolutions, STEM has overcome the fad threshold not only for its basis in sound learning theory, but also for a strong foundation on which the movement builds, and for the logical underpinning it relies on to infiltrate the curriculum, the school, the community, and society.

STEM Logic

For educators who have practiced the art and craft of teaching for more than a few years, it is not much of a challenge to check off a list of reforms that have come and gone. Many were good ideas in concept, such as brain-based learning, whole language, and differentiated instruction. Others were ill-defined forced fits such as discovery learning, new math, and "learning styles." All had the best of intentions. But for many reasons, sometimes institutional, often budgetary, and frequently lacking for evidence, they end up littering the educational roadside as crumpled fads. Schools are prone to fads, according to education scholar and commentator Larry Cuban in a classic April 2000 interview for *Educational Leader-*

ship (O'Neil 2000), because of all of the constituencies bringing the next big thing—school board members, vendors, a cadre of parents—rarely supported by evidence that one thing is superior to another. So a demeanor of "show me" befitting a Missourian suits teachers just fine as a bulwark against fads. Not only does evidence carry the day for veteran educators asked to adopt novel strategies, but most have a keen sense for the "eye test" when it comes to curriculum or pedagogy innova-

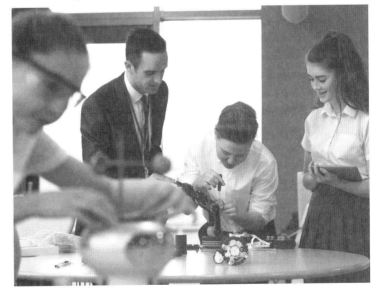

tion. Popularized by a committee charged with selecting teams for the College Football Championships recently, the *eye test* came to mean something immeasurable by statistics but discernable to the sports savant. Team X may be superior in all numeric dimensions but an expert's demotion of its rank was legitimized by witnessing unimpressive play, that is, the eye test. STEM passes teachers' eye test because it is sensible in the classroom and it is compelled by societal realities.

Early on in crafting a STEM agenda for Iowa, members of the Governor's STEM Advisory Council settled on a consensus definition furnished by Tsupros, Kohler, and Hallinen (2009):

> *[STEM is] an interdisciplinary approach to learning where rigorous academic concepts are coupled with real-world lessons as students apply science, technology, engineering and mathematics in contexts that make connections between school, community, work and the global enterprise enabling the development of STEM literacy and with it the ability to compete in the new economy.*

The sensibility of the definition, albeit slightly tautological, appealed first to educators on the Council. STEM by this measure passes the skeptical teacher's eye test—integrating of the disciplines, academic rigor, real-world applications, connecting school to work, and functionality in the world that awaits. Who could find fault with an educational innovation that holds such promise? But more than that, the logic threshold that filters fads from fixes reaches beyond classroom sensibility to community reality. And here, too, STEM passes the eye test.

Consider what awaits the class of 2020 and beyond. They are the generation that gets to define the bioethical boundaries for the revolutionary CRISPR (Clustered Regularly-Interspaced Short Palindromic Repeats) technology that empowers humans to tinker with the recipe for life—genes. *Science* magazine's *2015 Breakthrough of the Year* CRISPR are gene-editing tools used by biotechnologists to customize cells and organisms (Travis 2015). The potential medical and agricultural impact is dizzyingly profound—including the potential to "cure" diseases such as cystic fibrosis and sickle cell anemia (Trafton 2016). But ethical questions abound regarding "fixing" genes within reproductive cells and early embryos. Harvard University biomedical engineer and Des Moines Roosevelt High School alumnus Feng Zhang, a CRISPR researcher and front-runner for the Nobel Prize, observed recently that the great promise and peril of germ cell DNA editing calls for careful regu-

lation and "a lot more conversation" (Eller 2017). While they tackle CRISPR technology, our graduates will also be asked to develop and govern drones, artificial intelligence, social media metadata analytics, self-drive cars, and a host of emergent STEM controversies difficult to imagine now. Already the practical realities are out ahead of policy on unmanned aircraft systems (UAS, or drones). After years of drone proliferation, the U.S. Federal Aviation Administration imposed a December 21, 2015, and forward registration requirement, or risk committing a felony. The online registration site crashed twice in its first two days

under the tsunami of submitters, while policy regarding where and when UASs can be flown continues to evolve. Whether or not they find themselves developing these and other technologies or their policies, the class of 2020 and beyond will most certainly live better lives as consumers at the pharmacy or gas pump, as parents in the kitchen or at the pediatrician, and as citizens in the courtroom or the voting booth, if they enjoyed a K–12 STEM education. That is the other half of the eye test that commits teachers to STEM.

STEM's Moral Imperative

The STEM *awakening*, like a craft bazaar or a flea market, has something for everyone. Workforce and economic developers value the promise for creative and innovative talent for jobs of the future. Humanists value STEM education as a passport to meaningful professions accessible, ideally, by all. The business sector is STEM's most ardent driver at the local, regional, statewide, and national level currently, but a higher calling—the moral imperative for STEM—is nestled into the unalienable rights of the Declaration of Independence: life, liberty, and the pursuit of happiness.

Every American ought to be able to enjoy high-quality STEM education for the advantage it bestows its possessor regarding health, self-governance, problem solving, critical perspective on the news, and other perks of being educated. And if they so choose, every American ought to be able to pursue and thrive in a STEM career. Presently, that is not the case.

Currently underserved in STEM are students of minority race or ethnicity, including African American, Hispanic and Latino, American Indian, Alaska Native, and Pacific Islander. Two other categories of underserved are low income (combined parental income of $36,000 or less) and first generation to college. Nearly half of the 2014 ACT test takers who expressed interest in STEM were from one or more of these underserved groups. However, for students who self-identified as underrepresented minority, low income, and first generation college, there was a 34% drop in college readiness (ACT 2014). The enrollment trend for women in STEM has shown a steady increase of late, according to scholars Mack and McDermott at 46% from 30% in 2008, but the trend belies a persistent and troubling gender gap in physics, engineering, and computer science, as well as a precipitous drop-off of female enrollment at the graduate and professional levels of study across the STEM spectrum (2014). Conversely, the United States misses out on the talents of males disproportionately in the STEM disciplines of nursing, accounting, middle school teaching, and veterinary science (Elkins 2015). Also underserved according to their population proportion are the nation's learners who navigate disabilities of a broad spectrum spanning vision, speech, hearing, mobility, and learning. According to the National Science Foundation, persons with disabilities are underrepresented in the science and engineering workforce compared with the population as a whole (2013). Local and regional disproportionalities regarding access to high-quality STEM education sourced to geography (rural

versus urban for example), economy (community affluence), and opportunity (the difficult to employ such as the once-incarcerated or the underemployed) exacerbate STEM inequality.

That STEM education equips its recipient with skill sets to thrive in our technological 21st-century society both as a citizen and as a respectable wage earner makes it an unalienable right as a means to life, liberty, and the pursuit of happiness. Advocates and agents for STEM education for all, especially the underserved, carry the movement's highest calling and greatest promise: to uplift society. That is the moral imperative of STEM education.

STEM's Foundation

Among the many legends swirling about the origin of the STEM acronym, many credit Rita Colwell who, as director of the National Science Foundation from 1998 to 2004 is said to have determined that the agency's METS programs ought to be renamed. Over the course of the decade, STEM then evolved from merely linking four disciplines to representing an education movement that can revitalize the nation's economy. The previously mentioned report of the PCAST gave the movement a welcome boost with strong admonitions and the funding of programs to back them up. A goal put forth by the report was the following:

> To meet our needs for a STEM capable citizenry, a STEM proficient workforce, and future STEM experts, the Nation must focus on two complementary goals: We must prepare all students, including girls and minorities who are underrepresented in these fields, to be proficient in STEM subjects. And we must inspire all students to learn STEM and, in the process, motivate many of them to pursue STEM careers (PCAST 2010).

The year 2010 then became a tipping point for STEM in terms of reports and organizations hatching in support of the vision. Leading national voices Change the Equation, STEMconnector, and the NSTA's Task Force on STEM Education all launched in that timeframe. A flurry of seminal reports including the U.S. Chamber of Commerce's *The Case for Being Bold: A New Agenda for Business in Improving STEM Education* and National Governor's Association's Center for Best Practices in STEM report *Building a Science, Technology, Engineering, and Math Education Agenda* were published in 2011. The first-ever STEM Education Policy Conference was held by NSTA in Washington, D.C., in 2011. By May 2012, NSTA held the first STEM Expo and Forum focused on K–12 educator needs and solutions. Today, one would be hard-pressed to identify a nation, state, region, or locale void of STEM ambitions if not already marching toward the grand unifying goal for STEM put forth by the PCAST in 2010.

The flurry might lead one to believe that STEM's roots extend shallowly into just the 21st-century topsoil of the field of education. But like Newton who honorably recognized the shoulders of giants on which he stood to see further, STEM's rapid ascent has been possible only because of giant ideas decades old.

In 1982, NSTA issued a position paper on the topic of a reform effort in K–12 science education called *Science/Technology/Society* (STS). STS was anointed as a central goal for science education for the decade. Then in 1984, STS was branded a "megatrend" by the esteemed materials scientist Rustum Roy who went on to become a champion for the cause. The core mission of the STS approach was described as follows by Robert E. Yager, an early evaluator and subsequent advocate for STS and professor of Science Education at the University of Iowa:

> *[STS uses] … personal, societal, and career imperatives as organizers for schooling. STS is focusing upon real issues of today with the belief that working on them will require the concepts and processes so many consider basic. STS means starting with a situation—a question, problem, or issue—where a creative teacher can help students see the power and utility of basic concepts and processes. STS means starting with students, their questions, using all resources available to work for their resolution, and whenever possible, advancing to the stage of taking actual actions individually and in groups to resolve actual issues. STS makes science instruction current and a part of the real world (1996).*

If Yager's description of STS stirs recollection of the Tsupros team's definition for STEM it is not an accident or coincidence. "Career imperative," "real world," "all resources," "basic concepts," and "actual issues" were recognized decades ago as they are now to be lacking in many classrooms despite being favored in cognitive science circles—not to mention parent, child, and community circles. The *Next Generation Science Standards* (*NGSS*) include as Appendix J a helpful matrix summarizing how the interdependence of science, technology, engineering, and society can progress across the grade levels (NGSS Lead States 2013).

Instructive even today, though, is Yager's counsel of two counterforces inhibiting the widespread adoption of STS in schools: First, as a teaching philosophy more than a pre-packaged curriculum, STS had no off-the-shelf teaching manual or textbook to grab and go with. Second, a testing culture gripped education that overvalued basic concept recall at the cost of process skills and applications of knowledge. Yager's accumulated evidence of effect for STS teaching was seemingly undeniable: Students who experienced STS were better at applying science concepts, had better attitudes about science (especially female students), were more creative, held on to science information longer, understood the nature of science, and had greater

career awareness. Their teachers also became more confident about teaching science. Later in this book, Chapter 6 on learning and assessment in STEM will lay out a remarkably parallel array of effects for STEM teaching and learning.

STEM as descendent of STS begs the obvious question: In three decades, will we find that STEM—despite being an educational stress response spurred by a White House mandate, based on strong logic, passing the teachers' eye test, and built on an empirical foundation—still failed to enculturate, to merge with the DNA of schooling? The forces that inhibited STS (lack of packaged learning materials and a test addiction) are at work today on its progeny, STEM, but the difference is accumulated assets and allies that elevate STEM from merely a good idea if it can be fit in to an educational imperative.

STEM Elevators

What the STEM *awakening* (as opposed to a movement or revolution driven by some sort of popular uprising) has going for it are assets that amount to critical mass necessary to overcome the inertia of the status quo. Those assets include modern learning theory, new Standards, the proliferation of excellent off-the-shelf programs, and the support of the business sector.

Learning Theory

The *magnum opus* on what counts for effective teaching based on what we know about learning was delivered to humanity in 2000 by the Committee on Developments in the Science of Learning of the National Research Council. *How People Learn: Brain, Mind, Experience and School* removes any equivocation from and settles all debates regarding how learners need to be educated. Community engagement such as the formation of partnerships with business and the exploration of concepts in a context of utility are trademarks of the STEM learning experience and anchor effective teaching, according to the NRC committee:

> *Learning is influenced in fundamental ways by the context in which it takes place. A community-centered approach requires the development of norms for the classroom and school, as well as connections to the outside world, that support core learning values (Committee on Developments in the Science of Learning 2000, p. 25).*

The authors firmly rivet this point onto the educational psyche by quoting the visionary, would-be STEM advocate John Dewey from 1916:

> *From the standpoint of the child, the great waste in school comes from his inability to utilize the experience he gets outside, while on the other hand he is unable to apply in daily life what he is learning in school. That is the isolation of the school—its isolation from life (Committee on Developments in the Science of Learning 2000, p. 147).*

Dewey's goals for learning to result in utility and application are goals that resonate especially strongly in the STEM community. They rise to top billing as the NRC's ultimate goals for school:

> *[T]he ultimate goal of learning is to have access to information for a wide set of purposes—that the learning will in some way transfer to other circumstances. In this sense then, the ultimate goal of schooling is to help students transfer what they have learned in school to everyday settings of home, community, and workplace (Committee on Developments in the Science of Learning 2000, p. 73).*

Finally, learning is optimized when transfer and application-style growth experiences are coupled with multidisciplinary learning more characteristic of how challenges are encountered in the real world and which typify STEM classes. In-depth understanding is more likely "when subject matter domains overlap and share cognitive elements" (Committee on Developments in the Science of Learning 2000, p. 239).

New Standards

Carefully crafted to reflect the tenets of modern learning theory are three major assets for STEM advocacy and practice: the *Common Core State Standards for Mathematics* (*CCSS Mathematics*), the Common Career Technical Core Standards (CCTC), and the *Next Generation Science Standards* (*NGSS*).

The *CCSS Mathematics* were released in 2010, defining "the mathematics knowledge and skills students should gain throughout their K–12 education in order to graduate high school prepared to succeed in entry-level careers, introductory academic college courses, and workforce training programs" (NGAC and CCSSO 2010). Where STEM gets a big lift is in the eight Standards for Mathematical Practice that include abstract reasoning, argument construction, making sense of problems, and the critique of reasoning. Each supports a modern view of mathematics as more than mere formulaic procedural knowledge, but encompassing the skills and habits of mind that are logical job skills and life skills.

The *NGSS* were released in 2013, coming right out of the gate with a STEM-focused introduction: "The U.S. has a leaky K–12 science, technology, engineering and mathematics (STEM) talent pipeline, with too few students entering STEM majors and careers at every level—from those with relevant postsecondary certificates to PhD's. We need new science standards that stimulate and build interest in STEM" (NGSS Lead States 2013).

The standards are written as a series of performance objectives for science and engineering across disciplinary core ideas and crosscutting concepts ranging from optimizing design solutions to systems and system models.

Writers of the *NGSS* called out a number of "conceptual shifts" required for effectively implementing the *NGSS* as outlined in Appendix A of the standards. These shift advisories include the STEM-friendly notions that K–12 science education should reflect the real-world interconnections in science; that science and engineering are integrated in science education from K–12; that the *NGSS* are designed to prepare students for college, career, and citizenship; that standards should support broad interdisciplinarity; and that science standards coordinate with *CCSS for English Language Arts* and *CCSS Mathematics*.

The CCTC are divided into 16 Career Clusters that are subdivided into specific career pathways each with a set of standards. The set of STEM career clusters include competencies such as communication, creativity, critical thinking, and collaboration (Advance CTE 2017). Another key set of standards for technology provide strong support and guidance for STEM educators. Both the International Technology and Engineering Educators Association (ITEEA) and the International Society for Technology Education (ISTE) bolster the cause with such contributory guidelines as "Evaluate and select information sources and digital tools based on the appropriateness to different tasks" (ISTE 2016); and "Students will develop an understanding of Design. This includes knowing about the role of troubleshooting, research and development, invention and innovation, and experimentation in problem solving" (International Technology Education Association 2007).

A compelling case for enculturating STEM rests on learning theory and the standards that incorporate its tenets. Two more modern-day assets less prominent in the STS era, which put STEM on firm footing, are off-the-shelf programs and the unprecedented support of the business sector.

Rise of Pre-Packaged STEM Programs

STEM educators and their community partners have at their disposal a luxuriant array of high-quality programs. Many align strongly to the aforementioned standards, some are robustly evaluated for efficacy, and most can pass the discerning teacher's eye test. There are programs that introduce elementary youth to engineering such as the Museum of Science, Boston's, *Engineering is Elementary* and Project Lead The Way's *Launch*. There are programs that integrate STEM learning into the excitement of auto racing or flight such as *Ten80 Racing Challenge* and *Real World Design Challenge*, respectively. Robotics can be a highly successful STEM entry point through programs such as *FIRST* (For Inspiration and Recognition of Science and Technology) and *BEST* (Boosting Engineering, Science and Technology). For learning environments lending more to online platforms, *Learning Blade* and *Defined STEM* offer customized content, process, and career exploration for K–12 students. Finally, think LinkedIn for students who would benefit by connecting to STEM professionals, and programs such as *Nepris* and *STEM Premiere* come to mind.

This is merely a sampling of programs with which Iowa has enjoyed some experience. The national organization Change the Equation maintains a rigorously vetted list of exemplary programs in its STEMworks Database where a more extensive roster with far more detail is maintained (*http://changetheequation.org/stemworks*).

The nonprofits and businesses that have made an investment in packaging exemplary STEM teaching and learning products are contributing significantly to the crystallization of STEM into American school culture.

Backing of the Business Sector

Speaking of investments in STEM, the business sector has rallied in unprecedented fashion to devote time, talents, and treasures to enact a call by the U.S. Chamber of Commerce for bold, ambitious reforms to STEM education in its Institute for a Competitive Workforce report of 2011 titled appropriately *The Case for Being Bold*. "Business leaders are equipped to provide the kind of straight-talking leadership and relevant expertise that transformative STEM reform requires," states the report, calling on business leaders to "focus on specific key areas: taking full advantage of strengthened and streamlined academic standards; rethinking how teachers are hired, deployed, and prepared; and promoting new models of schooling that can facilitate STEM learning" (Hess, Kelly, and Meeks 2011). The commitment of dollars to STEM education across this nation on the part of the business community is significant. Boeing announced

in 2015 for example a $15 million K–12 STEM program (Feldman and Longacre 2015). Intel the same year, as described in the piece, "The Business Case for STEM Education" in *Fortune* magazine, announced a $300 million K–12 and college STEM investment to promote talent diversity (Lev-Ram 2015). And at the White House Science Fair in 2015, then-President Obama announced more than $240 million in private-sector pledges to inspire and prepare more youth to excel in the STEM fields. That is a significant percentage of the $2.9 billion for STEM education that the president wrote into to his 2015 federal budget. Future federal investment in STEM education is less clear.

Specific businesses and their classroom-level investments in STEM will be the subject of Chapter 4. In recognition of the human capital production role of K–12 and college STEM education, the private sector is all-in providing a hill-topping impetus for

advancement. There is a lot in STEM for the business community to like, as explored in the following section.

Why STEM Matters to a Business Leader

Mike Ralston is the president of the Iowa Association of Business and Industry (ABI), a diverse group of 1,500 businesses, including manufacturers, retailers, insurance companies, financial institutions, publishers and printers, transportation services, law firms, and health care organizations large and small, urban and rural, together employing more than 330,000 Iowans. Many ABI members as well as the organization itself are strong champions and partners for STEM education. ABI launched its own careers initiative, *Elevate Advanced Manufacturing*, to strengthen the talent pipeline through education and marketing. Ralston took some time to expound on the value of STEM for the business sector.

"Business people are excited about the STEM movement for two reason. First, because they believe it will result in more—and more capable—employees. Second, because they

believe it is simply good for Iowa to have this important matter receive attention" (personal communication). Ralston and the sector for whom he speaks are quick to remind STEM leaders of the importance of data and evidence of progress. Such metrics as STEM career interest and abilities among high school graduates, post-secondary enrollment trends, diversity of enrollees, and employment trends in the STEM fields are tracked and reported annually to demonstrate headway. He considers such communication the most vital part of ensuring continued support and overall success of STEM.

Ralston does not believe there is a favorite or most important intervention among the suite of solutions that make up local, regional, or statewide STEM programs. They may range from teacher professional development to camps and after-school programs for youth, and from public awareness campaigns to specialized STEM schools. Reform efforts can stretch into preschools and up to college and beyond. Ralston observes that the business sector is supportive of each strategy and the entire spectrum of beneficiaries. Some private-sector partners prefer to invest in early childhood STEM programming in deference to research supporting the establishment of a strong foundation. Others opt in at the middle or secondary school levels with more college and career-ready initiatives. And yet other business partners prefer to collaborate at the postsecondary level with internships and recruitment fairs. Many engage at all of these levels as well as connecting to communities through STEM events, competitions, and fairs.

The voice of business was captured through a 2011 study of the McKinsey Global Institute titled *An Economy that Works: Job Creation and America's Future*. It reinforces a concern that the education system respond to an increasingly dire workforce condition by noting that enrollment trends portend that in the next decade the United States will produce twice as many graduates in the social sciences and business as in science, technology, engineering, and mathematics. If nothing is done to redirect the interests of some of those graduates toward STEM, expect an exacerbated shortage of qualified candidates for technical jobs (Manyika et al. 2011). Already 64% of companies surveyed for the study report a lack of qualified applicants especially in science and computing areas. Naturally, business supporting STEM education is a matter of self-preservation and economic viability in the years to come.

Why STEM Matters to a School Administrator

John Carver leads a rural Iowa school district, Howard-Winneshiek Community Schools, a vast region of farms and small towns nestled in the gently rolling hills of the Northeast corner of the state. Despite immense budget challenges linked to enrollment decline resulting from urban migration typical across Midwest states, Carver has shepherded his schools, families, and communities squarely into the 21st century through a 1:1 program by which all K–12 students connect with the world beyond their school walls and learn to think globally. His district is an Apple Distinguished Program for Innovation, Leadership and Educational Excellence and is featured in the *2016 National Technology Education Plan* of the U.S. Department of Education. Carver is a founding member of the Iowa Governor's STEM Advisory Council.

"Chuck Yeager was the first man with the courage to take the technology of that day, the Bell X-1 (1947), and move through extreme turbulence to break the sound barrier" said Carver in setting up an explanation for why STEM is swooping up school administrators.[1] "For STEM to be fully deployed and operational means taking the technology we have today and summoning up the courage to navigate through the extreme turbulence of resistance to change to break [our own version of] the sound barrier." Carver then segued to another technology metaphor to make his point, "We are at a 'printing press' moment in the history of mankind. Just as the printing press made books available to the masses, facilitating the exchange of thinking and accelerating creation of knowledge, digital devices connected to the internet are doing the same at never before experienced speed. Change now happens not in years or days, but almost instantly, in micro moments." Carver's attentiveness to the fleeting window of time available to educators for making their mark with students is exemplified by a brand rollout in his district. The 2020 Howard Winn program sets a goal for district graduates to be second to none in preparedness for success by the time current sixth graders walk the stage. To reach that goal, his district makes STEM a priority.

"In the 20th-century industrial age, reading, writing, and arithmetic were the fundamental survival skills to employability. Emerging skill sets for employability and being able

---

1 Quotes in this section are from the author's personal communications with Carver in December 2015.

to contribute to society in the 21st century go well beyond the 3*R*s. Now more than ever, aptitude and ability in Science, Technology, Engineering, and Math (STEM) will define and determine 21st-century success," according to Carver, while reflectively adding the importance of coupling literacy with STEM.

Carver's district works closely with the private sector of the region in building meaningful learning opportunities. For administrators navigating STEM partnerships, he cautions that education systems, classrooms, schools, and districts, have operated in isolated silos, insulated from the world of commerce where seemingly different language is spoken and different goals are in play. "Historically when business and education have talked it has been education asking for money or business coming into the school for 'one shot' career day" observes Carver, "Transparent relationships need to be built based on identifying common touch points." Ultimately it is economic development interest that unites schools and businesses. Education leaders who envision their work as prominently including human capital production for the world of work make valuable business partners.

The internal dynamic of the school district is managed simultaneously by administrators embracing STEM. Carver has succeeded in building a culture of STEM by making a case for change. "The organization needs to build urgency for change," according to Carver, boiling it down to a basic truism, "If the organization does not see or desire change, it will not happen." But that change, he advises, is more likely if the new priority is integrated. "The pitfall one must avoid is thinking of STEM as 'bolt on' or aftermarket 'special' activities. Like reading, STEM must be embedded into learning." But Carver is adamant that no agenda can be forced. STEM imposed on a team not fully bought in "is like putting a size 10 shoe onto a size 8 foot. It will be sloppy, unattractive, and uncomfortable and the wearer will discard the shoe and go barefoot." When a culture of STEM sets in though, the telltale signs Carver has come to recognize include empowered teachers and students moving from consumers of content to creators of content, thousands of followers from across the country and around the world on the district's Facebook site, and learning taking place 24-7, not just during school time.

Why STEM Matters to a Teacher

The proverbial rubber meets the road when teachers match-make students and STEM in the classroom. If reading and writing were once the great equalizers that provided a passport to a good living for any American willing to work at it, today it is STEM skills such as coding, statistics, logic, reasoning, critical thinking, creativity, collaboration, and

problem solving. No one understands that better than Shelly Vanyo, science teacher and department head at Boone High School in North Central Iowa, and Kacia Cain, biotechnology and physiology teacher at Des Moines' Central Campus, a regional academic center for central Iowa. Cain also serves on the Governor's STEM Advisory Council. Vanyo and Cain both received the 2015 STEM Education Award for Inspired Teaching sponsored by Kemin Industries in recognition of their initiative in forming local business partnerships and developing unique learning experiences for students.

Vanyo gravitated to STEM as an alternative to the prevalent "set and get" classroom culture. "I changed my focus from what students needed to know that day/week/semester to how to best prepare them to be global citizens," said Vanyo, altering her students' and her own conception of her classroom to be an "innovation zone" less driven to correct answers, and more about the journey and broader outcomes—skills and big ideas.[2] She puts a premium on real-life applications of classroom learning and elicits mighty effort of her students by building trusting relationships. Failure is a powerful teaching tool that Vanyo leverages to inspire her students to heights they may not have known they were capable of. Her advice to other teachers dabbling in but yet to fully immerse in STEM is to "try relating an objective to something students personally enjoy. For instance, to help students learn about gas laws, I invited a parent who pilots hot air balloons to teach lesson on the gas laws related to burners." Similarly, Cain weaves real-life application into everything asked of her students. "By providing my students with opportunities to interact with both industry and higher education researchers, I am giving them 21st-century skills in the content area of science." And like Vanyo, her goals for students lie well beyond the immediate: "Classroom experiences in inquiry provide them with the foundation to be successful beyond high school."

The STEM fires that burn in Vanyo are fueled by strong, positive role models and mentors with whom she surrounds herself, from within and beyond the school. Cain thrives on "watching my students' confidence levels soar as they master different skills, techniques, and information that set them apart from their peers and gives them a competitive edge at the next level." Cain's tactics that provide that edge feature research projects collaboratively with universities and nearby industries that allow students to connect with the next level and see themselves as biological professionals.

School and community culture for STEM is important for the professional well-being of both Cain and Vanyo. The director of Cain's Central Campus, Aiddy Phomvisay, weighed in on how a culture for STEM is maintained, "on the premise that student learning and engagement should be meaningful and applicable immediately to their lives, future careers, and the world around them." And that manifests, according to Phomvisay, in all of their programs from automotive technologies where "students are using micron meters to measure and evaluate disc brakes" to marine biology and aquarium science applying biochemistry to maintaining healthy water ecosystems, to "our CSI Criminal Justice lab whereby students collect, secure, and analyze human DNA from a mock homicide crime scene."

---

2 Quotes in this section are from the author's personal communications with Vanyo and Cain in December 2015.

Creating a STEM Culture for Teaching and Learning

Central Campus has succeeded in making STEM "not just an acronym or slogan but a way of learning and doing."

When Vanyo considers the broader culture of her school and community, she buys in to the importance of perceptions. "Sometimes teachers and parents perceive an active, exciting classroom environment as not serious and lacking success," and so she takes perceptions head-on, "I invite administrators and fellow teachers into my classroom as 'expert advisers' and send students home to engage their parents as learning partners in kitchen table inquiry." And what signs tell Shelly that she's moved past the tipping point with parents and students especially? "Instead of communicating about grades [with parents], we communicate about skills, progression, future plans." And as for students, "They no longer ask 'Is this the right answer?' and instead enthusiastically proclaim 'Look what I found!'"

The Story From Here

In the chapters that follow, the starry ambitions of STEM are brought down to Earth in the pragmatic realities of schools and communities. Building on the foundational principles

of STEM and the *edu-nomic* rally that unifies the public and private sectors around schooling, Chapter 2 presents the essential role of networks—intra- and inter-school and cross-stakeholder alliances— for effective, sustainable STEM education. The journalist's principle of the Five *W*s leave no stone unturned regarding networks: Who are the essential players? What are the roles and responsibilities? When do networks function well? Where are the boundaries and proximities that characterize effective STEM networks? Why is networking important?

If a network can be envisioned as the orchestra bringing STEM dreams to life, then a community would be the audience. Chapter 3 portrays the community as historically neglected in educational innovation yet serving as the linchpin to "viral" expansion of the good habits, practices, and goals of STEM. Schools and organizations that capitalize on the power of public awareness strategies win over

audiences. Revisiting the notion that citizens' perceptions of school become their reality, successful STEM programs are at work shaping public perceptions.

Then comes a hallmark of STEM—the school–business connection. What was once the sole domain of career-technical education, workplace applications of content and on-the-job learning permeate the core disciplines of math and science in a STEM school culture. Chapter 4 characterizes thriving models that bridge the widening education–industry chasm, defining the roles of teachers, students, and business partners. The emergence of work-based learning with its potency in soft skills/employability skills/success skills/human skills development holds great promise.

Next, we consider how early 21st-century STEM has enjoyed a bandwagon effect as the brand strengthens. Purists and pioneers have put structures in place to establish a certain standard that, in Chapter 5, upholds important labels such as STEM teacher, STEM classroom, and STEM school. These structures, standards, and their effect are explored.

Interwoven with classroom and teacher designations are curricula and assessment strategies. Chapter 6 profiles some of the exemplary materials that support STEM teaching and learning, as well as the emergent process of homegrown school–business learning platforms. Investigative, interdisciplinary, and applied learning as defines STEM according to Tsupros, Kohler, and Hallinen (2009) mean a flipping of assessment—both what is assessed and how. Vanguard thinkers in this space are creatively dovetailing local and state-mandated assessments with authentic measures.

There are teachers of STEM subjects and then there are STEM teachers. Chapter 7 distinguishes the two, highlighting preparatory programs and pathways that integrate the best practices illuminated up to this point into the minting of a professional STEM educator. Of all the dimensions of STEM education reform, new teacher production may be the most delicate front for its institutional constraints and market forces. That makes it a most important front as well, where innovators are taking on the challenge.

Much as driverless cars and metadata technologies have jumped out ahead of regulation policy, STEM education leapt out front of the support systems that aid consistency and ensure high quality. Nowhere is this more evident than the hit-and-miss landscape of teacher professional development in STEM. Chapter 8 examines the challenges inherent in helping content area secondary teachers and generalist elementary teachers implement a STEM learning mission. Outstanding models are poised for scaling.

Finally, despite a 30-year on-ramp to the interstate STEM roadway, a number of hazards remain to be navigated. Equally important are the gains to be celebrated as STEM fundamentally transforms education systems. Chapter 9 is the result of answered invitations for experts across the STEM stakeholder spectrum to weigh in on where we go from here. Classroom practitioners, industry advocates, community catalysts, parents, elected officials, and interested bystanders all offer unique threads of thought that together form a fabric of the future of STEM.

References

ACT. 2014. *Understanding the underserved learner.* Iowa City: ACT. *www.act.org/content/dam/act/unsecured/documents/STEM-Underserved-Learner.pdf.*

ACT. 2015. *Condition of college and career readiness report 2015.* Iowa City: ACT. *www.act.org/content/act/en/research/condition-of-college-and-career-readiness-report-2015.html?page=0&chapter=0.*

Advance CTE. 2017. Common career technical core. *https://careertech.org/CCTC.*

Committee on Developments in the Science of Learning. 2000. *How people learn: Brain, mind, experience and school.* Washington, DC: National Academies Press.

Elkins, K. *Business Insider.* 2015. 20 Jobs That Are Dominated by Women. February 17.

Eller, D. *Des Moines Register.* 2017. A Roosevelt Alum Describes What Gene Editing Has in Common With 'Twinkle, Twinkle, Little Star.' March 22.

Feldman, D., and J. Longacre. 2015. Boeing announces $15 million investment in children and STEM learning. Boeing. *www.boeing.com/company/about-bca/washington/boeing-announces-15-million-investment-in-children-and-stem-learning-07-24-2015.page.*

Goldstein, D. *New York Times.* 2017. Obama Education Rules Are Swept Aside by Congress. March 9.

Hess, F., A. Kelly, and O. Meeks. 2011. The case for being bold: A new agenda for business in improving STEM education. Washington, DC: Institute for a Competitive Workforce, U.S. Chamber of Commerce. *www.uschamberfoundation.org/sites/default/files/publication/edu/The%20Case%20for%20Being%20Bold_2011_v2_0.pdf.*

International Society for Technology Education (ISTE). 2016. ISTE standards for students. *www.iste.org/standards/standards/for-students-2016.*

International Technology Education Association (ISTE). 2007. Standards for technology literacy. *www.iteea.org/File.aspx?id=67767&v=b26b7852.*

Lev-Ram, M. *Fortune.* 2015. The Business Case for STEM Education. January 22.

Mack, K., and P. McDermott. 2014. The twenty-first century case for inclusive excellence in STEM. *Peer Review* 16 (2): 4–6.

Manyika, J., L. Lund, B. Auguste, L. Mendonca, T. Welsh, and S. Ramaswamy. 2011. An economy that works: Job creation and America's future. McKinsey Global Institute. *www.mckinsey.com/global-themes/employment-and-growth/an-economy-that-works-for-us-job-creation.*

National Governors Association Center for Best Practices and Council of Chief State School Officers (NGAC and CCSSO). 2010. *Common core state standards.* Washington, DC: NGAC and CCSSO.

National Science Foundation. 2013. Women, minorities, and persons with disabilities in science and engineering. *www.nsf.gov/statistics/wmpd/2013.*

NGSS Lead States. 2013. *Next Generation Science Standards: For states, by states.* Washington, DC: National Academies Press. *www.nextgenscience.org/next-generation-science-standards.*

O'Neil, J. 2000. Fads and fireflies: The difficulties of sustaining change. *Educational Leadership* 57 (7): 6–9.

President's Council of Advisors on Science and Technology (PCAST). 2010. *Prepare and inspire: K–12 education in science, technology, engineering, and math (STEM) for America's future.* Washington, DC: Executive Office of the President.

Society for Human Resource Management (SHRM). 2006. *Are they really ready to work? Employers' perspectives on the basic knowledge and applied skills of new entrants to the 21st century U.S. workforce.* Alexandria, VA: SHRM. *www.p21.org/storage/documents/FINAL_REPORT_PDF09-29-06.pdf.*

Stein, J. *Time.* 2013. Millennials: The Me Me Me Generation. May 20.

Trafton, A. *MIT News.* 2016. Curing Disease by Repairing Faulty Genes: New Delivery Method Boosts Efficiency of CRISPR Genome-Editing System. February 1.

Travis, J. *Science.* 2015. Making the cut CRISPR genome-editing technology shows its power. December 17.

Tsupros, N., R. Kohler, and J. Hallinen. 2009. STEM education in Southwestern Pennsylvania: Report of a project to identify the missing components. Intermediate Unit 1: Center for STEM Education and Leonard Gelfand Center for Service Learning and Outreach, Carnegie Mellon University, Pittsburgh, Pennsylvania. *www.cmu.edu/gelfand/documents/stem-survey-report-cmu-iu1.pdf.*

Yager, R. E. 1996. *Science/technology/society as reform in science education.* Albany, NY: SUNY Press.

CHAPTER 2
Catalyzing Professional STEM Networks
Local, Regional, and Statewide

Impactful and sustained STEM calls for alliances. Like a super-saturated solution in need of just a small disturbance to crystallize, community advocates require a common cause, a convening to crystallize into a STEM network. Iowa's STEM professional network is shown in Figure 2.1 (p. 24).

Figure 2.1. Iowa's STEM Professional Network

Iowans (dots) and their connections (lines) grow and intertwine over time through the state's STEM network.

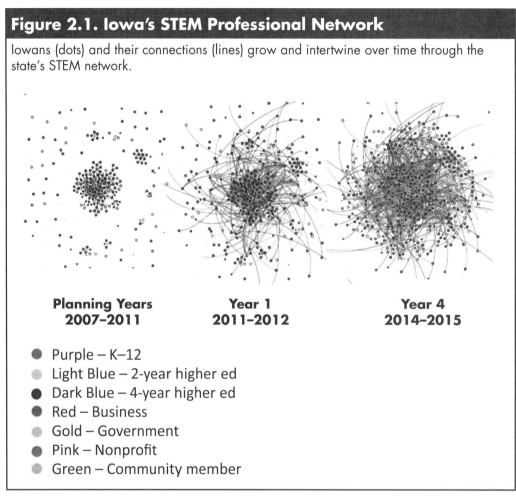

Planning Years
2007–2011

Year 1
2011–2012

Year 4
2014–2015

- Purple – K–12
- Light Blue – 2-year higher ed
- Dark Blue – 4-year higher ed
- Red – Business
- Gold – Government
- Pink – Nonprofit
- Green – Community member

Source: Mari Kemis and Andres Lopez, Iowa State University. National Science Foundation grant number DRL-1238211.

Note: See a full-color version of Figure 2.1 on the inside back cover of this book.

Networks Are About Resources and Communication

For any disruptive innovation to scale, a network is needed. Consider the disruption to a pasture when termites invade. A single colony consisting of a million workers, soldiers, kings, and queens efficiently distributes resources across hundreds of chambers through chemical communication and specialized roles. Is it so much of a stretch to liken termite networks to the Roman Empire? Wrapped around the Mediterranean Sea at its heyday in the 1st century AD, the Roman Empire spanned more than 3 million miles and about 70 million people. Provinces were ruled by governors dispatched from Rome who carried news and extracted resources (taxes and goods). Very disruptive initially, the Empire

endured through tumult for a thousand years because of its distributed, networked organization. Are the networks of termites and Romans relatable to modern social network systems *sine qua non*—through the internet? Successfully disrupting the lodging industry is all about leveraging a network to scale, which is AirBnB's quintessence. Doing business in 191 countries listing 3 million places to sleep in 65,000 cities, AirBnB scales by connecting ever-more hosts with guests, basically distributing resources through communications—the recurrent tenets for networks (AirBnB 2017). The same is true for Uber, Lyft, and other transportation-related networks as well as social media such as Twitter and Facebook, not to mention shopping sites like eBay and Amazon. When you boil it down, networks emerged where there had been none, and the function is essentially resource distribution (a ride, a status update, a book) through communication.

America's STEM education *awakening* is no different. Networks are driving the advance. Like so many termites, Roman soldiers, or Uber drivers, advocates for an interdisciplinary, applied, collaborative, problem-solving approach to learning disrupt the status quo. And they can do it only by networking. The feedback cycle that strengthens networks comes alive when members realize that together they can know more and do more without costing more. That epiphany was captured in the moment by a rural school superintendent speaking at a statewide education conference when asked why he bothers with inter-district, multicommunity partnerships. "Collaboration and cooperation have allowed

us to do more with limited resources," said Dan Cox, superintendent of the Charles City, Iowa, school district who counters declining enrollment and therefore state appropriation through resource distribution and communication (i.e., networking).[1]

The remainder of this chapter examines STEM networks through two lenses. First, the journalist's five *W*s are applied to networks: what, why, who, when, and where. Second, networks through the eyes of practitioners at the local, regional, state, and national levels reveal the secret sauce behind the disruptive educational innovation that is STEM.

---

1 Cox stated this during his remarks at the Iowa STEM School+Business Innovation conference on June 29, 2016.

WHAT Are Networks?

The rich and fascinating field of sociocultural evolution is built entirely on the premise that *Homo sapiens* is a species genetically predisposed to networking. It is a Darwinian fundamental: Humans who communicate and distribute resources across social circles better weather calamities and dodge predators and as a result, pass on the genes for such capacities. Thus for the purpose of STEM education, networking is an innate tendency undeniable in just about everybody to collaborate. A value-add to socializing is that new ideas are born in mixed company. Einstein once said, "What a person does on his own, without being stimulated by the thoughts and experiences of others, is even in the best of cases rather paltry and monotonous." Networks are incubators of innovation. Good ideas are not only more likely in mixed company but also more likely to spread. Sociologist Ron Burt (1995), in his book *Structural Holes: The Social Structure of Competition*, describes studying hundreds of business managers who, depending on how connected they were across diverse organizations, generated more and better ideas. The spread of good ideas has its own branch of science supporting a theory of Diffusion of Innovations that strives to elucidate how and why some ideas go "viral" while others wither.

In his landmark book named for the theory, Everett Rogers positioned communication across social networks as pivotal to the spread of an innovation. His four classifications of "adopters" of a new idea are familiar to anyone in the innovation market—there are first the innovators, then the early adopters, then the early majority, followed by the late majority, and finally the laggards (2003). Somewhere near the junction from early to late majority (about 50% of any population, naturally) an idea or invention hits a tipping point where it spreads across the culture (e.g., AirBnB) or begins its death spiral (e.g., New Coke or the XFL).

STEM networks spread the wealth of knowledge and generate new ideas for education. If a network can be envisioned as the orchestra bringing STEM dreams to life, then

a community would be the audience. Except this would be that type of show where the musicians descend from the stage to march up the aisles and engage the audience, drawing everyone into singing and dancing along. America's STEM awakening is likely to be in Rogers's late, early majority phase of diffusing innovation, approaching a tipping point of cultural infusion. But many sublevels of STEM networks may exist at all points on

the continuum. STEM networks may be intra-organizational (a singular school or school district), local (community partnerships), regional (a geographic alliance, for example the Midwest STEM Forum of central U.S. states), and of course statewide (a number of states, including California, Iowa, Massachusetts, Ohio, South Carolina, Texas, and many more). Nationally, STEM networks have been established by nonprofits (Change the Equation and Project Lead The Way, for example) and trade organizations (e.g., the Aerospace Industries Association). Stories of advantageous networks at each level are discussed next.

WHY Network?

Faster-Better-Cheaper (FBC) was NASA's mantra before the Columbia shuttle disaster in 2003 when closer examination determined the mutual exclusivity of the terms. Pick any two, said analysts upon fully dissecting the tragedy, but you cannot have all three. A Google search of FBC yields a million hits, attesting to the philosophical and practical wrangling over its tenability. Under what conditions can costs be reduced while quality improves and production time shrinks? Possibly only by networking.

Case in point: At Vermeer Corporation, a global manufacturer of heavy-duty agriculture and industrial equipment based in Pella, Iowa, Faster-Better-Cheaper can be realized under special circumstances that involve efficiency and communication, according to the company's Director of Continuous Improvement Gary Coppock. By clearly communicating high expectations to suppliers early and often, Vermeer Corporation is able to reduce the waste that accompanies trial and error. Its network connects activities within a value stream and eliminates waste in communication and processes resulting in better flow of information and materials. The elimination of waste leads to faster decisions, ordering, and delivery. Networks are all about communication and resources, and in the case of a global manufacturer, it enables faster delivery, higher quality, and controlled cost of production. In other words, faster, better, and cheaper (in terms of cost, not quality) with no shortcuts.[2]

As a result of the better ideas that arise from social circles, combined with the efficiency gained by connecting, as exemplified at Vermeer, networks can clearly provide a competitive advantage. In their wonderful book *The Innovator's DNA*, Dyer, Gregersen, and Christensen highlight this magical power of networks whereby participants often stumble onto unforeseen benefits. A nutritionist entrepreneur discovers a Malaysian fruit of healing folklore and leverages his network to build a $1 billion company around it. A backyard

---

2 This discussion is based on personal communications between the author and Coppock in April 2016.

barbeque couples a microbiologist with a manufacturer who builds a bioremediation company around pollution-eating bacteria. A high-tech ceramics start-up struggles with quality variance until connecting with a photographic films expert who had solved its chemical mix challenge decades ago (2011). What's good for business is even better for STEM education—multidisciplinary conversations, inter-organizational cross-pollination, and the out of the box thinking that results from wide and diverse networks serves STEM well as an engine of innovation for education.

Boiling the advantage of networks down to numbers, a paper in the esteemed journal *Nature* called "Quantifying Social Group Evolution" by Hungarian scientists Palla, Barabási, and Vicsek (2007) shone a bright light of empiricism on human tendencies to network. The authors developed a tidy algorithm from studying the origins and durations of collaborations that arise within and across professional communities—a science of its own, charmingly labeled *clique percolation*. It turns out that the healthiest of large networks constantly adapt and evolve by drawing new members whose fresh perspectives and ever-changing needs alter the community's direction in real time, keeping the network relevant and cutting edge. Interestingly by contrast, small networks endure through stability of a membership composition that remains pretty much unchanged over time. But emergent small networks, or percolating cliques, can thrive in the duality of stability forged through member tenure, and freshness brought by new faces and ideas, through linking the local, stable network to the larger, evolving one. Many examples of such dualities surface among the practitioner profiles to follow.

WHO Makes Up a STEM Network?

At the risk of overusing the team metaphor, networks are perfect analogues. STEM network teams come into being when one or a few influencers decide to get in the game, triggered by a desire to communicate and share resources. The scale of engagement may be the community-level leagues, championed by a school administrator or local business leader. Or, regional visionaries such as legislators or major employers come to recognize how much they could learn and grow by connecting across regions. State-level influencers including agency heads, trade organization directors, and often governors help launch inter-state leagues, or networks. These are the

catalysts for whom this chapter is named. Regardless of the scale, all successful networks begin with and are sustained by one or more catalyst. These influencers who form teams and join leagues are a curious sort commonly driven by a conviction that together is better, and that when it comes to STEM, a window of opportunity commands action. Their sports analogue would be the team owner or general manager, without whom there is no league.

Rarely do the influencers make the unilateral decision to start a STEM team or join a network. The awareness of and desire to improve STEM education among executive-level policy leaders, school superintendents, higher education officials, and corporate CEOs is an indispensable constant in the STEM network equation, but to get things done they rely on experts in the field. The experts are coaches—professionals in the STEM space who have experience as practitioners as well as managers. They embody a blend of street credibility and organizational savvy that earns the allegiance of implementers who, for culture shift to take hold, need to join in at a level of critical mass or tipping point.

In his book, *The Tipping Point*, Malcolm Gladwell (2002) presents three rules for any trend to reach a tipping point where it stitches into the cultural fabric:

1. Trends start with an influential few who spread an idea.

2. The idea endures or sticks because it is different, it stands out, and it has an appeal.

3. The trend becomes the new normal when a sufficient number of key implementers at the local level adopt it, swaying the late adopters and laggards (to co-mingle Gladwell with Rogers' theory of Diffusion of Innovations).

The implementers are the performers themselves, players on the court or field, and educators in the STEM arena. Teachers, day-care providers, scout masters, librarians, job trainers, robotics team sponsors, newsmakers, church leaders, camp counselors, and curriculum vendors are some of the key players who network through organizers responding to the opportunity presented by influencers.

What sports metaphor would be complete without mention of the fans? The *essentialia* of any network team is to serve its constituency. Serena Williams can pack a stadium by showcasing unrivaled grace and athleticism, and unbridled joy for the game of tennis. Her contagious and compelling persona inspires the next generation of *raquetteurs*. She is thereby serving a fundamental goal of the womens' professional tennis network, a goal even more basic than that of making money—Williams helps the network sustain itself by drawing young fans and future talent. STEM networks serve a constituency of learners and their families. The value of those learners to the network is the promise they embody for sustaining or invigorating local, state, or national economies. As stated in Chapter 1, the STEM *awakening* is a stress response to the growing perception that education and workforce are out of sync such that graduates may not be queued up for modern jobs. The STEM awakening is cheered on by employers, school boards, mayors, and everyone else dependent on a steady talent stream for the relatively

CHAPTER 2

recession-proof, egalitarian, well-paying, and rewarding careers of STEM. The fans served by STEM networks are right now somewhere between cradle and career (or diaper and diploma, or toddler and tassel, or maybe ga-ga and graduate).

WHEN to Network?

Timing is everything for both comedians and STEM networkers. Too early and one can burn lots of time and energy trying to win over an influencer or coaches and implementers who are not yet ready. Too late, and the community can fragment or get territorial. In his book *When: The Art of Perfect Timing*, Stuart Albert (2013) details a vast array of analytics that guide the logic of leaders who make critically timed decisions. Almost all of them are acquired by playing the game of chess. He advises that one consider the sequence of events that need to unfold before acting, the rate of acting such that events do not unfold too swiftly or slowly, and the risks one is ready to accept by acting. A STEM network catalyst times her outreach to coincide with the establishment of prerequisite conditions. Is there sufficient awareness of STEM? Is there a commonly held view of a problem that STEM can fix (i.e., work-ready graduates)? Do key decision makers see active roles for themselves in the STEM fix? Under such conditions, networking is more likely to be successful. But

even when all of the prerequisites for success align, there will be *the dark night of the innovator*, a phrase linked to Hewlett Packard in the 1990s that has become common parlance in the anatomy-of-innovation sphere (Figure 2.2). It lies somewhere ahead of Gladwell's (2002) tipping point where rate and risk keep network catalysts awake in the wee hours. Is the network's formation moving too slowly such that early joiners drift on account of impatience or perceptions that the network is making no difference? Or conversely, is the network expanding and progressing at a pace that makes significant numbers of members uneasy? Does the formation and growth of a STEM network risk ground gained on other education or workforce development efforts by siphoning off energies and talents? Youthful as the STEM awakening is, its roadside is already littered with the carcasses of dead networks that sputtered under sequence, rate, and risk miscalculations. But more often than not, fortunately, conditions are ripe for STEM networking, and numerous examples hum along under the leadership of strategic catalysts.

Figure 2.2. Innovation Anatomy

A familiar state, the dark night of the innovator, afflicts STEM network pioneers midway from launch to pay-off, as framed by the anatomy of innovation.

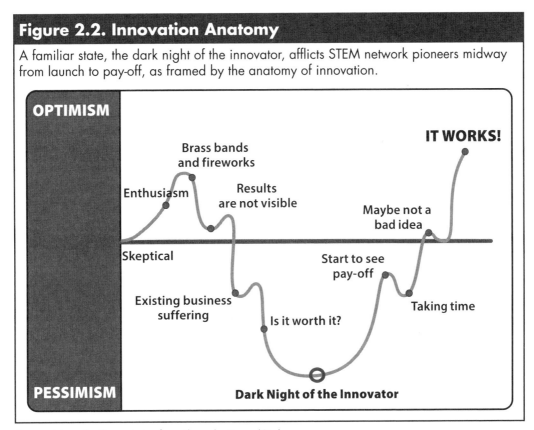

Source: Adapted with permission from Alexandra Drane/@adrane.

WHERE Are Networks?

The ubiquity of STEM networks attest to an American culture shift toward an education system that embraces interdisciplinarity, collaborative problem solving, and community connections. Like many U.S. states, Iowa has become highly networked in the STEM sector as depicted in Figure 2.3 (p. 32). The out-of-school community of scout troops, daycares, libraries, zoos, museums, and after-school clubs has become unprecedentedly connected under the STEM banner. Individual schools, entire districts, and regional school consortia have established STEM networks. Partnerships between higher education, industry, nonprofits, trade organizations, and preK–12 have sprung up. An organizational structure created by the Governor's STEM Advisory Council in 2011 has the state carved into six STEM regions each administered by a Hub housing a STEM manager. One key role for the managers is to catalyze additional local and regional STEM networks (see *www.iowastem.gov/regions* for more information).

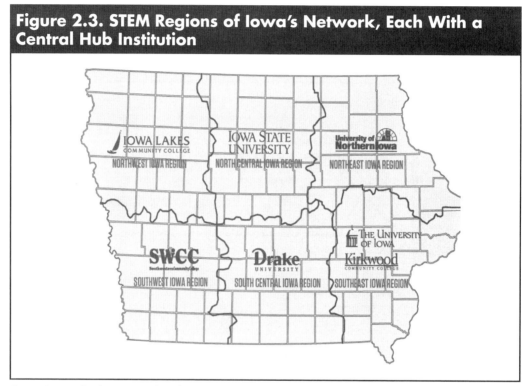

Figure 2.3. STEM Regions of Iowa's Network, Each With a Central Hub Institution

Source: Reprinted with permission from the Iowa Governor's STEM Advisory Council.

The organization STEMconnector keeps a list of state and regional STEM networks at *www.stemconnector.org/state-by-state.* A sampling of the array, ranging from well-established to dormant to emergent, and from well-funded to getting by on a shoestring, include the following (in alphabetical order):

- California, *www.cslnet.org*
- Georgia, *http://stemgeorgia.org*
- Indiana, *www.istemnetwork.org*
- Maine, *http://mainestem.org/stem-collaborative*
- Massachusetts, *www.mass.edu/stem*
- Minnesota, *http://scimathmn.org/mnstemnet*
- New York, *www.nysstemeducation.org*
- North Carolina, *www.ncstemcenter.org*
- Ohio, *www.osln.org*
- South Carolina, *www.s2temsc.org*

- Tennessee, *http://thetsin.org*

- Utah, *http://stem.utah.gov*

- Washington, *www.washingtonstem.org*

- Wisconsin, *www.wistem.org*

Interstate and regional networks layer atop state efforts as well. For example, a Midwest STEM consortium of 13 states convenes annually to compare playbooks and share best practices. The Noyce Foundation funds a cost-free STEM Ecosystems Initiative that connects dozens of cities and regions across the country with the goal of supporting STEM learning through sharing and technical assistance (*http://stemecosystems.org/first-community-of-practice*). Some states have joined together under the banner of the Ohio-based research firm Battelle to form the inter-state STEMx network, a fee-based collaborative (*www.stemx.us*). Many U.S. states have joined an interstate fee-based STEM Consortium hosted by the International Technology and Engineering Association (ITEEA) to share and develop curriculum and other resources (*www.iteea.org/12/389.aspx*). Finally, nonprofits such as FIRST (For Inspiration and Recognition of Science and Technology, *www.firstinspires.org*) and Project Lead The Way (*www.pltw.org*) as well as industry groups such as the Aerospace Industries Association (*www.aia-aerospace.org*) and the Society of Automotive Engineers (*www.sae.org*) coordinate nationwide and global STEM networks.

What follow are profiles of five successful STEM networks ranging from the local (intra-school), to community, to region, to state, to national scope.

Window Into STEM Networks

Intra-school STEM Network: Saint Theresa Catholic School

Saint Theresa Catholic School in Des Moines, Iowa, has STEM fever. It was destiny, given the leadership of Principal Ellen Stemler. She has seen to the STEM professional development of more than 80% of her staff so that the school's 300 students enjoy a broad and deep suite of experiences ranging from computer programming to green energy gadgetry, robotics, food chemistry, and more.

St. Theresa's intra-school STEM network sprung from a core group of teachers committed to owning the STEM brand in fulfillment of its vision for inspired learners capable of pursuing STEM dreams. By now the network has mushroomed to ensnare the School Board, the Parent Teacher Club, individual parents, engineering mentors from the parish and school community, alumni, community businesses, nearby universities, and even elected officials.

The visioning process that launched St. Theresa's distinguished STEM record started years ago. "We were focused on improving the learning for our students and providing 21st-century learning opportunities for them," recalled Stemler. "We researched STEM

learning, met with a parent with a research science background and then looked at STEM programming available through the Governor's STEM Advisory Council."[3] What followed were a cavalcade of trainings and resource acquisitions to equip every teacher in the school for wall-to-wall, bell-to-bell STEM integration. She stands sentry for anything that can help the school's mission. "We take advantage of every STEM opportunity that is available," including "the state's STEM conference, workshops at area universities, competitions for students, externships and training for teachers, and any chance for students to share the depth of their learning." To power the school's STEM emphasis has required some financial rethinking. "We have made budgetary changes including adding a STEM component to our school foundation," explained Stemler. Her team's resourcefulness has led to investments by local companies, parents, and community organizations. The parish to which the school belongs has declared STEM to be a funding priority.

Communication of ambitions figured prominently in the growth of St. Theresa's STEM network (foretelling the upshot of Chapter 3—effective messaging). The school website and print/online publications, School Board and Parent Teacher Club presentations, the church bulletin, and presentations to community groups helped get the word out, and to steel the brand. A feedback cycle of benefits followed, according to Stemler, "The more we share at the community level the more the community wants to partner with us."

When it comes to advice to would-be networkers at the intra-school level, Stemler recommends a preliminary assessment of existent and potential partnerships. "Who has something to share that will benefit your students?" This is the time to take advantage of resources for the good of students. She continues, "Make a list of your current partnerships and those you would like to develop. Be strategic in your planning." And if at first you don't succeed, "Keep communicating so that the businesses can see your successes. It may take several visits and conversations before a potential networker says yes."

Upon forging STEM partnerships, Stemler prioritizes communication. "Invite the community into your school," and between showcases, she advises empowering students to tell

3 Quotes in this section are from the author's personal communications with Stemler in May 2016.

the story: "Pictures really do speak a thousand words. Have students share their learning with their own words in videos, surveys, brochures and pictures."

When the school's STEM intra-network is humming, Stemler sees the reward as "independent, resilient students who want to solve problems." For detailed information about Saint Theresa Catholic School, visit *http://sainttheresaiowa.org/school*.

Community STEM Network: Rocket Manufacturing

Rocket Manufacturing is not a NASA-funded missile maker, but a high school custom-order shop working with local industries. Almost six years in the making, Rocket Manufacturing exemplifies community networking to shape rather than await the future. Tucked in the rural Northwest corner of Iowa, the Rock Valley Community School District graduates about 50 students each year. Thanks to visionary leaders who bridged the education–industry chasm, those students have the opportunity to develop marketable skilled trades and work habits en route to their diplomas.

Rock Valley Superintendent Chad Janzen recalled the genesis of Rocket Manufacturing back in 2008, when a local business leader shared a magazine story about a Wisconsin school housing a student-run manufacturing business. One road-trip and school board presentation later, plans for a 10,000 square-foot shop began to emerge. Rocket Manufacturing was to be a public-private partnership built on a basic question that educators posed to business leaders out of the gates: "What are you looking for in a good employee?" It has been a positive relationship of mutual benefit ever since.

Rock Valley may do more manufacturing per capita than any community on Earth. There are an estimated 1,200 metal fabricators (machinists, welders, laser cutters, etc.) at work in the vicinity of the town, population 3,000. Employers struggle mightily to fill jobs. Thus, Rocket Manufacturing met a willing ear among an essential network of professionals including local economic developers, business executives, community college leaders, and vitally, school teachers and administrators. Students at the school now crank out custom metal products on contract to area manufacturers. And the benefits of the partnership, according to Janzen, include more than just technical skills, but also the all-important employability skills like collaboration and responsibility. "Students are empowered to achieve things they didn't realize they were capable of," he said, noting the biggest reward of operating a profitable business in the school is the "self-confident students that are success communicators, problem-solvers, and producers."[4]

Janzen's advice for others pondering community STEM partnerships is to identify a business "cheerleader." "You need a person or two in your community to support what you are trying to do," he coaches, adding, "partnerships must be seen as mutually beneficial while the focus must be on the kids." The Rocket Manufacturing operation is detailed at *http://rocketmanufacturing.weebly.com*.

---

4 This discussion is based on personal communications between the author and Janzen in May 2016.

Regional STEM Network: Quad City Engineering and Science Council

The Quad City Engineering and Science Council (QCESC) spans four cities in two states, bridging the Illinois communities of Moline and Rock Island to the Iowa cities of Davenport and Bettendorf and vicinity. To paraphrase an old Barbara Mandrell country ballad, they were STEM before STEM was cool. Launched in the early 1960s as a nonprofit umbrella organization, the QCESC is made up of about 25 engineering, science, and technical societies in the Quad City area with approximately 3,000 associated professional members. Those members range from practicing engineers and scientists to formal and informal educators (preK–16), local government, nonprofits, parents, and students.

Born of a necessity for uniting a number of local technical societies to more effectively publicize the importance of engineering and science to the advancement of communities and quality of life, the mission of the QCESC has evolved to facilitate "increased collaboration and support for STEM education and supporting the next generation of innovators for students in preK–12 through local business, education and nonprofit organizational partnerships," according to Pat Barnes, the organization's executive director emeritus and program director for John Deere's Global STEM initiative *Inspire*. The long-term sustainability of the QCESC, according to Barnes, is "a strong, passionate group of volunteers who work well together and have a shared vision of supporting current and future STEM professionals."[5]

The services provided by the QCESC include information exchange through the website and social media regarding events (e.g., Annual National Engineers Week Banquet), professional development (e.g., STEM Teachers Night Out), and news. The other hallmark of STEM networks, resource access, is provided by the QCESC in the form of awards (e.g., Engineer and Scientist of the Year and STEM Teacher and Volunteer of the Year Award), scholarships for college STEM majors, and volunteer coordination.

At more than 55 years old, the QCESC is a thriving organization supporting more than 20 STEM-related community events each year, requiring more than 400 volunteers from various companies, organizations, and schools, affecting more than 3,000 people of whom

---

5 This discussion is based on personal communications between the author and Barnes in June 2016.

2,000 are preK–12 students of the area. Their number of regional robotics tournament teams has increased tenfold in the last decade. The number of scholarships per year has reached 16 to the tune of $36,000. And in 2016, 278 students from 14 area high schools took part in its Quad City Tech Challenge.

Sustainability for the QCESC hinges on a familiar factor for any network advocate: cycling leadership. Especially as demand for its services continues to grow, the key challenge ahead, according to Barnes, is "the number of volunteers we can actively engage. This includes replacing current key board members and volunteers as they retire with a similar passion to current supporters." Meeting that challenge means attracting new organizations and businesses to the group.

The secret to building a successful regional STEM network, according to Barnes, is to "create an open, very inviting network where everyone is welcome and information is freely shared." Strategically connecting instrumental people and organizations has kept the QCESC fresh. "Formation of key partnerships includes retirees, local and global businesses, local organizations that support kids (Girl Scouts, Boy Scouts, 4H, Home Schoolers), parents, and professionals," says Barnes, "who believe that together we can do more for the next generation of STEM professionals." Details about the Quad City Engineering and Science Council may be accessed at *www.qcesc.org/index.html.*

State STEM Network: South Carolina

A number of U.S. states have enacted STEM networks but few have been at it as long as South Carolina. According to Dr. Tom Peters, executive director of South Carolina's Coalition for Mathematics and Science (SCCMS), parent organization to the state's network of S²TEM (Solutions in Science, Technology, Engineering and Mathematics) centers, it dates back more than two decades to a grant from the National Science Foundation.[6] "We had NSF support for 10 years—two rounds of five years each," said Peters, recalling that as many as 26 states were awarded Statewide Systemic Initiative (SSI) grants in the 1990s. South Carolina was one of eight states to receive a second round of SSI funding. Today, Peters knows a thing or two about sustainability. He explains, "I believe we are the only state maintaining the network formed through the SSI."

The South Carolina STEM network endures, indeed thrives where others have stalled, "by growing the STEM capabilities of others," Peters reveals. He captured common threads woven through Barnes's regional network, Janzen's community network, Stemler's

---

6 The following discussion is based on personal communications between the author and Peters in June 2016.

intra-school network, as well as every other successful network profiled in this chapter by illuminating the three ways that the SCCMS grows STEM capabilities:

1. Support learners with information, innovations, and research.
2. Grow leaders by identifying and developing their skills.
3. Leverage resources by aligning ideas and efforts.

Ever since the era of NSF support, South Carolina's STEM network mission has broadened while the resource base has diminished. "We have adopted a much more 'business like' approach to managing our network and to the delivery of programs and services," observed Peters, "Our advocacy network has broadened with our mission and our funding base has diversified significantly." Today the SCCMS supports learners through services such as writing the state's standards in mathematics and science and conducting regional STEM festivals. It grows leaders by preparing and supporting instructional coaches and offering online courses. And it leverages resources by aligning efforts through state and national summits and partnerships with corporate entities including BMW, DuPont, Michelin North America, Progress Energy, and the South Carolina Department of Education. And all the while, the SCCMS underpins programs and services with research and evaluation for data-driven visioning.

The enduring maturation of South Carolina's STEM network boils down to adaptability, versatility, and accessibility. "Whereas once we were all about teachers and schools, we are now also about communities and children," explained Peters regarding mission adaptability. He casts the organization's financial adaptability in this light: "Whereas once we were funded almost exclusively by NSF and then by the State of South Carolina, we are now funded by grants from a multitude of sources, gifts from business/industry, fees for services, and some assistance from our General Assembly, too."

"Versatility," interprets Peters, "[is our] ability to see ahead to new challenges and consistency in the quality of the support we provide to our constituents. And by accessibility, Peters's sentiment aligns well with the science of *clique percolation* through strategic growth. "We manage lots of brands now that all interconnect with our mission and with each other," he noted the tactical strength, "I think it is critical to have many access points into the STEM ecosystem." For details on South Carolina's Coalition for Mathematics and Science, visit *www.sccoalition.org*.

National STEM Network: Change the Equation

The rapidity with which STEM has taken hold in schools, communities, regions, and states across the United States caused a momentary national network leadership vacuum filled beautifully by Change the Equation (CTEq). In November 2009, the White House launched the Educate to Innovate initiative intended "to move American students from the middle to the top of the pack in science and math achievement over the next decade." Among the flurry

of tactics unfurled was the recruitment of more than 100 major company CEOs in 2010 to launch CTEq, "a new nonprofit with full-time staff dedicated to mobilizing the business community to improve the quality of STEM education in the United States" (*https://obama whitehouse.archives.gov/issues/education/k–12/educate-innovate*).

Today, the CTEq network has diversified to include not only business leaders but also STEM education program providers and state-level STEM champions. Connections with state STEM leaders has become particularly important according to Claus von Zastrow, chief operating officer and director of research, "We try hard to maintain a strong network of state STEM state champions, especially as the Every Student Succeeds Act shifts power back to the states."

But the backbone of CTEq's network are partner companies. They had joined up "out of a concern for the nation's economic competitiveness at a time when most of the general public did not know what STEM stood for" said von Zastrow. Their aim in joining forces was to close the United States' STEM skills gap through communication and resource enhancement—functions now signature at this point in the chapter for any useful network—namely by doing the following:

- Highlighting the growing opportunities for those with strong STEM skills
- Offering clear and actionable diagnoses of the condition of STEM education in every state
- Advocating for common-sense solutions to state- and national-level challenges to improving STEM education
- Improving the impact of companies' engagement in STEM education by increasing the return on their investment in STEM education

In six years, CTEq has figured prominently in the national movement to make STEM a household word. "We don't have to include the parenthetical '(Science, Technology, Engineering, and Math)' anymore," Claus noted as evidence of progress, "and the fact that strong STEM skills open doors seems widely accepted."[7] Leading though not solely contributing to awareness gains, the CTEq network's research and materials have been widely used

7 This discussion is based on personal communications between the author and von Zastrow in June 2016.

by national and state advocates who continue to press the case for STEM. And in the last two years CTEq's mission has expanded with state and corporate partners to identify and promulgate best practices through its STEMworks program, affecting roughly one million more children from 2014 to 2015. Considerable growth in program scope is expected in the years to come.

Network capacity for CTEq has upscaled significantly as a result of the STEMworks honor roll of effective STEM programs, bringing, as von Zastrow frames it, "a network of leading STEM education advocates and program providers who have on-the-ground experience of what works in STEM education." The organization has also carefully culti-vated ties to state STEM organizations in a symbiosis of material, information, and human resource interdependency. "These state and local partners understand where the needs for STEM intervention are greatest, what political dynamics can help or hinder our shared work, and what other partners should be at the table with us," von Zastrow observed, believing that state partners alongside corporate members "can help us collectively reach hundreds of thousands of U.S. youth with the nation's best STEM education opportunities."

Sustaining the CTEq network, von Zastrow, like South Carolina's Tom Peters, aligns philosophically with the empirical derivations of Palla, Barabási, and Vicsek (2007) dis-cussed earlier: Successful networks maintain various entry points and continuously draw new partners to the effort, by "understanding that new partners can invigorate our shared work, and by understanding that credit for achievements has to be shared fully by all the members of the network," as von Zastrow succinctly puts it. A challenge to long-term thinking of STEM networks, according to von Zastrow, is the "big political and eco-nomic developments continually changing the cast of characters in companies and state government, so the work of building and maintaining a coalition never ends."

Advice from von Zastrow regarding networking centers on drawing in the influencers "who can provide political cover and financial fuel," he said, explaining that "I've seen far too many well-meaning networks flounder, because no one in the network has a strong connection to key decision makers or funders in a state or a community." And far more than

an afterthought, von Zastrow emphasizes a commitment to high-quality and evaluation, "Base your strategies on research into what works, and continually measure the impact of what you do." For more information about CTEq visit *www.changetheequation.org*.

"Innetworkovation"

At a STEM conference[8] Patricia Elizondo, an executive at Xerox Corporation, said this about disrupting the status quo: "Innovation can be hard, and risky, and very lonely, like being the first dancer on the floor. But when you get something right, now you've got a conga line." STEM networks are conga lines that relieve and bolster innovators to scale. Networks are so vital to broad scale innovation that they ought to be one word: *innetworkovation*. Networkers like rural school superintendent Dan Cox can do more with less resources. Einstein branded life paltry and monotonous without networks. Vermeer Manufacturing can make more product, better, and of higher quality. Saint Theresa Catholic School transformed the entire culture. The QCESC's inter-city incubator develops talent. South Carolina and CTEq adapt and grow missions of vast impact. All of them reflect those aspects of master networkers who opened this discussion—termites, the Roman Empire, AirBnB—dancing alone at first perhaps, but now leading conga lines. STEM advances on the shoulders of *innetworkovators*.

References

AirBnB. 2017. About us. *www.airbnb.com/about/about-us*.

Burt, R. 1995. *Structural holes: The social structure of competition.* Cambridge, MA: Harvard University Press.

Dyer, J., H. Gregersen, and C. Christensen. 2011. *The innovator's DNA: Mastering the five skills of disruptive innovators.* Boston: Harvard Business Review.

Gladwell, M. 2002. *The tipping point: How little things can make a big difference.* New York: Back Bay Books.

Palla, G., A.-L. Barabási, and T. Vicsek. 2007. Quantifying social group evolution. *Nature* 446 (7136): 664–667.

Rogers, E. 2003. *Diffusion of innovations.* New York: Free Press.

Stuart, A. 2013. *When: The art of perfect timing.* San Francisco: Jossey-Bass.

8 National Alliance for Partnerships in Equity Professional Development Institute, Arlington, Virginia, April 17, 2012.

CHAPTER 3

Community Buy-in for STEM

It is at the community level where STEM dreams come to life. But the community in recent history has been neglected in educational innovation, despite serving as the linchpin to "viral" expansion of the good habits, practices, and goals of STEM. Of all the great possibilities in cross-pollination from the world of business to education, messaging ranks near the top. Schools and organizations that capitalize on the power of public awareness strategies win over audiences. Revisiting the notion that citizens' perceptions of school become their reality, successful STEM programs are at work shaping public perceptions.

Re-entry of the Community to Educational Discourse

The marrow of any STEM network skeleton is the community, where the magic happens. Local, regional, statewide, and national networks are re-engaging communities in

the discourse as well as delivering new era educational programming. This is a recent resurgence. An historic educational upheaval equal to today's STEM disruption predates it by almost a century. As so many upheavals go, it was an innovation made possible by a new technology that held great promise but ushered an unforeseen disconnect between schools and their communities. The innovation was regional comprehensive schools, made possible by a new technology, the school bus.

The school bus was a supremely disruptive technological innovation for 20th-century education.

School and Community Decoupled

Blue Bird, the brand burned into the psyche of every transported school child, turned out its first steel bus in 1927. Built on the chassis of a Model T, it could carry a dozen youth at speeds up to 40 miles per hour at about 20 miles per gallon. With a tank of gas

costing about $2.50, lots of kids could be moved a few miles each day for very little money. And that helped spell the demise of the one-room school.

An American education was once a quite personal, umbilical connection of learners to their communities. At their peak in the late 19th century, Iowa had more than 12,000 one-room schools, within walking distance for nearly anyone seeking to learn. The schools were built by the local citizenry who recruited the teacher, collectively contributed to salary, and rotated responsibility for the teacher's room and board. The curriculum asked basic skills development in reading, writing, and arithmetic—essential know-how to function and contribute as farmers and merchants. The era of an agrarian economy linked schools and their communities locally and viscerally. Yet, pressures were looming—forecast by electrification, telephones and radio, cars, trucks, highways, banking, and manufacturing—on schools. Functionality in the rural community was inadequate for what Peter Drucker (1992) coined the term *Knowledge Economy* around the turn of the 20th century.

A new set of basic facts and skills provided the graduate a visa to a career in industry (and increasingly, access to higher education). To set the table for a more modern curriculum, the National Education Association's "Committee of Ten" in 1892 rolled out expectations that persist to this day—classes in abstract mathematics, the suite of sciences including biology, chemistry, and physics, as well as history and foreign language. It is in that confluence of higher expectations, specialized learning, and school buses that community connections to schooling frayed and popped loose. Youth were trucked off to regional schools, divided into grade levels, and taught the universal curriculum of success-for-the-times by imported subject matter experts. Everyone graduated with at least a command of long division, stoichiometry, and Hamlet. The product of K–12 was factory job-ready or queued up for college.

Divergent Evolution of Schools and Communities

Drucker's Knowledge Economy gave way to the Information Age with the rise of computers and the internet in the late 20th century. The need for subject matter experts to pour information into the heads of learners evaporated. "Today knowledge is free. It's like air, it's like water," said Tony Wagner of Harvard University's Innovation Lab, at the 2015 Second Annual Symposium on Global STEM Education at Harvard Graduate School of Education. He continued, "There's no competitive advantage to knowing more than the person next to you. The world doesn't care what you know. What the world cares about is what you can do with what you know."

Meanwhile it has been said that a time traveler up from the 19th century would find nearly everything about our era bewildering except maybe the instructional mode of some of our schools and colleges. By contrast, a tour through a John Deere tractor factory—with its computer-numeric–controlled robotic assemblers, precision axle milling to the micron, triangulated satellite guided autopilot control systems—would render our time tourist catatonic. The bewildering pace of advancement in the modern workplace coupled with our lost competitive advantage for knowing lots of things are the factors driving a reconnection of schools and their communities. Schools cannot create the modern product—collaborative, adaptive, real-world problem solvers—without help from the community. And Eric Engelmann knows why.

Engelmann is the president of a software company, Geonetric, in Cedar Rapids, Iowa. They specialize in web-based health-care tools. He revealed at the 2014 Association of Business and Industry's Taking Care of Business conference what it is that keeps him up at night, "My 70 employees who are more like family depend on me and this business to stay fresh and front-edge because right now somewhere in Paris or Tokyo or Tucson somebody

is building a competitive product. We've got to be constantly striving, improving, and innovating." Geonetric, like all global businesses (except maybe a rare few who occupy such a specialized niche that they have no competitors … yet) dwell in a piranha pit of accelerated Darwinian evolution. New technologies and strategies that build efficiency and profit are the traits of fitness for survival. The advent of Kaizen-like practices toward lean and continuous improvement are examples of strategies that spread through the pit enhancing the fitness/profitability of early adopters. Cybersecurity of proprietary information is an essential technological adaptation acquired by survivors in the competitive piranha pit of global business. Corporations must constantly evolve, sometimes dramatically rapidly, as did McDonald's with an all-day breakfast, and Microsoft with the acquisition of LinkedIn to access a vast professional network, or perish like Pets.com, RadioShack, Blockbuster, Borders Books, Circuit City, Atari, Polaroid, and so on.

Public schools, by contrast, rarely dwell in such competitive environments (though the charter movement, vouchers, and online schools are dialing up the heat in some regions)—more of a bucolic pond of meandering koi and catfish than a pit of piranhas. Schools evolve, too, of course, but at a comparatively leisurely pace. New technologies are adopted—smart boards, 1:1 digital devices, learning diagnostics—but their impact on learning can be incremental more than revolutionary. Teachers take part in professional development to differentiate instruction, flip the classroom, and bring gaming to learning, with each innovation moving the needle, not breaking the gauge. Over time, varied survival pressures result in almost unrecognizable relatives—schools and the communities that depend on them.

A natural analogue to the school–community disconnect is dolphin–hippo divergence. They branched off the same family lineage about 50 million years ago, during one of Earth's warm eras where seas were expansive and land was swampier thanks to melted polar ice caps. One side of the family tree opted for the carnivorous life and hit the water in pursuit of small fish, the other chose swamp life and vegetarianism. A 2009 research study points out similarities in their ears, skin, bones, and teeth not to mention genes, which solidly link the lineages of hippos and dolphins. But time and environment can really do a number on appearances. Bulky legs and deep-set ear parts were more costly than helpful underwater. And a trachea venting the lungs out of forward-positioned nostrils simply would not do. A watery world favored ear drums right up on the skin, and fins where legs were. A blowhole atop the head that can exchange air in a fifth of a second was the answer to nostrils undersea. Millennia and survival adaptations separated two close relatives (Spaulding, O'Leary, and Gatesy 2009).

Time and environment have also separated the world of education from its community of employers and workplaces that await graduates. Environmental factors including busing, teaching for a knowledge economy, and the onset of the information age (now on to the Innovation Age, according to Tarak Modi's insightful 2011 book *Living in the Innovation Age*), compounded by time, has resulted in a gap—the two sides hardly recognize one another. That gap is an opportunity space where STEM enters in (Figure 3.1).

Figure 3.1. The STEM Opportunity Gap

The arithmetic pace of change in schools: Steady improvements to infrastructure and professional development means a better experience each year (left). The logarithmic pace of change in business: Rapid improvement through research and development, adaptive processes, and acquisitions (right) create an opportunity gap for STEM.

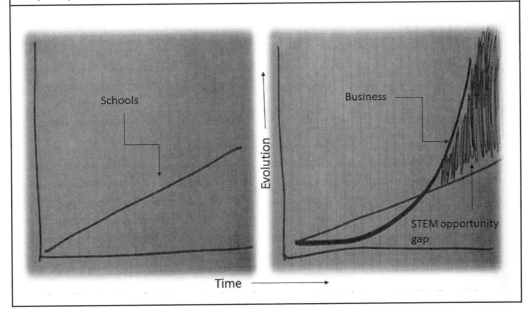

School and Community Coupled Again

STEM connects the bucolic educational pond of koi and catfish to the piranha pit of the broader community, making both better. Consider STEM a channel introducing cutting-edge business practices to education while awakening the work world to its talent pipeline—wonders, warts, and all. As a hybrid reform movement, STEM has been called an *edu-nomic* development initiative, a term that generates over a million Google search hits (through some sort of algorithmic miracle, Iowa STEM usage shows up in position 7—or positions 1, 6, and 8 with *development* attached—as of the time of this writing). A quick aside is in order, an homage to career and technical education (CTE): school–business partnerships are the heartbeat of CTE and have been since the birth of the nation (remember Johnny Tremain?). STEM brings such partnerships mainstream in terms of both lateral connections across core disciplines, and horizontally kindergarten through grade 12, as will be explored in more detail in the next chapter. How schools are re-engaging communities in the name of STEM may take many forms, with three key ways dominating the landscape: school–business collaborations, messaging, and bringing STEM to the community.

School–Business Collaborations

The breadth of this topic and its explosive growth as both a priority and practice in schools across the country merits a chapter solely devoted to exploring a thorough profile. For now, the imperative that has brought business to the table to lend a hand in talent production was succinctly stated by Andreas Schleicher, director of education and skills at the Organization for Economic Cooperation and Development, who said at the Governor's 2016 Future Ready Iowa Summit, "Your school system today is your economy of tomorrow." The community has come to grips with that reality and in some places, brings the cavalry to help at school. More to come in Chapter 4, devoted to school–business partnerships.

Messaging

The field of education in general, and STEM education particularly for our purposes, could take a lesson from the engineering profession. In 2008, the National Academy of Engineering produced a report *Changing the Conversation: Messages for Improving Public Understanding of Engineering*. The report had two overarching goals: to elevate the importance of the profession in the minds of the public, and to draw more youth into the profession. The mechanisms developed include tactical messages and strategies for getting the word out. Taglines that emerged from the study included "Engineers make a world of difference," and "Engineers help shape the future." Professional societies, academic institutions, and corporate partners were charged with carrying the message forward (National Academy of Engineering Committee on Public Understanding of Engineering Messages 2008). Today, engineering is one of the fastest growing undergraduate majors, on a steady incline since 2007 including a 7.5% enrollment spike in 2014–2015 according to the American Society for Engineering Education working paper "Engineering by the Numbers" (Yoder 2015). Engineering colleges across the country are experiencing enrollments boosts, including Michigan State, Olin College, the University of Oklahoma, Belmont, Youngstown State, MIT, and Iowa State, to name but a few.

STEM education could be the best thing for U.S. education since Horace Mann, but if the community does not get the memo, all of the evidence and expertise is stunted. Yet messaging is often an afterthought, or at least undervalued. When Iowa launched a statewide STEM initiative in 2011, a public awareness campaign was proposed by members of the Governor's STEM Advisory Council. The rationale centered on citizen buy-in, which would steel legislators' resolve. But a counterargument prevailed, that precious taxpayer resources ought to be maximally committed to programming directly benefiting youth rather than to a seemingly nebulous ad campaign aimed at manipulating lawmakers and the public. It was a principled stance that tested well with constituents. And it was wrong.

Among the two-dozen metrics annually gauged by evaluators of the state's STEM program is public awareness and support of STEM. A random survey of a sufficient number of Iowans to permit generalization to all the citizens of the state shows a doubling (from 26%

to more than 50%) who recognize the acronym STEM since 2012 (Heiden et al. 2016). And respondents affirm the STEM imperative at a 90% clip once familiarized. Public service announcements on television and radio, billboards along the interstates, social media posts, tweets, pins and chats, bumper stickers, lapel pins, newspaper columns, branded trinkets, event exhibits, and more tactics carry Iowa's STEM message to the masses today as a result of a grand compromise that vaulted public awareness from ill-advised to indispensable. Leaders agreed to a communications budget line of about 3% of the legislature's STEM appropriation so long as vendor bids included a dollar-for-dollar match. The brand "Greatness STEMs From Iowans" debuted in 2013 and has never looked back (Figure 3.2). The campaign proved to be a cyclic feedback loop of awareness-begetting support that figured prominently in achieving a tipping point of engagement across agencies and organizations joining the STEM awakening.

The Design of a State's STEM-Messaging Campaign

In 2013, the advertising agency Strategic America (SA) of West Des Moines, Iowa, was selected from a competitive pool of applicants to develop the state's STEM brand and message. SA works with such brands as Lennox, Wendy's, Pella Windows and Doors, Toro, and other national as well as local clients. Matching the state dollar-for-dollar, SA focuses on media relations, events, and a messaging toolkit to drive the Greatness STEMs From Iowans campaign.

Branding is the primary tool that SA wields to drive excitement and inspire action around any product or service. A bright, invigorating logo of fun colors is high up on the early to-do list. The team at SA came up with the Greatness STEMs From Iowans logo and look (Figure 3.2) to "Convey the idea that greatness can come from the young people of Iowa." Courtney Shaw, STEM account manager for SA, revealed the inner workings of creative minds as "A play on words, visual appeal and contemporary block type intentionally tilted [to] indicate that this is not your ordinary teaching platform, but something fun and exciting."

Once branded, the message was shaped in collaboration with leaders of the state STEM initiative. A goal was forged from a molten stew of factors ranging from academic skills to economics, from public enthusiasm to career connections. SA arrived at a mission to "create a public awareness campaign that increases interest in and awareness of STEM education in Iowa, and develop a creative platform that will incite ingenuity, deliver key messages and build on the existing efforts and activities of the STEM

Figure 3.2. Iowa STEM Logo

Brand can contribute significantly to public support for STEM. The advertising agency Strategic America produced this logo for Iowa STEM.

initiative." Target audience came next, which were "key supporters in accepting and promoting the STEM initiative" according to Shaw and included primarily leadership and members of the Governor's STEM Advisory Council, followed by K–12, youth agencies and nonprofits, higher education, employers, parents, legislators, and media outlets themselves.

The strategies and tactics employed by SA to execute Iowa's STEM integrated communications campaign focused on four main areas: public and media relations, events, billboards and public service announcements, and a messaging toolkit. When it came to media relations Shaw detailed, "Our team created and maintained a statewide media list to pitch unique story ideas to outlets across the state. We also hosted media training sessions, coordinated the launch of the Iowa STEM PSA, created talking points, wrote opinion guest editorials, coordinated speaking engagements, created media packets, and conducted various pitches resulting in media coverage."

SA also helped the STEM Council with major events, including STEM Day at the Iowa State Fair, the Lt. Governor's STEM Town Hall Tour, and others. STEM billboards dotted the state, and a television commercial featuring Pinterest CEO and Iowa native Ben Silbermann was broadcast statewide. Finally, a toolkit that included a message map, PowerPoint presentations, lapel pins, letterheads, business cards, banners, bookmarks, brochures, flyers, Greatness STEMs From Iowans video, and other items was delivered to the Council and various STEM supporters.

Measuring the effect of SA's awareness campaign is important to STEM leaders, especially in answer to the early (and lingering) skeptics. Findings from the aforementioned public survey are critical feedback that equip SA "to create measureable goals that align with the direction of Iowa's STEM initiative" according to Shaw. In addition to data from STEM's annual survey, the SA team also conducts quarterly media coverage reports detailing the number of media placements, impressions, locations, key messages, and open rates for all STEM stories published across the state over time. All the while, she keeps focused on understanding the wants, needs, and desires of the target audience, resulting in a substantial increase in the interest and awareness of STEM in Iowa. And a value-add unanticipated by the STEM Council but figuring prominently in the partnership's success is that the agency really cares. "Our devotion to this initiative shines through every single project we have done, and will do," Shaw affirmed, "because we are passionate about STEM."

Bringing STEM to the Community

This chapter rounds out with a third and most powerful mechanism by which schools are re-engaging communities in the name of STEM: bringing STEM to the community. Citizen access to the community school is often limited to sporting events, band concerts, parent-teacher conferences, and, for a limited few, school board seats. Otherwise it can be something of a black-box mystery what takes place between 8 a.m. and 3 p.m. behind those walls, save for the school day revelations that may spill forth over the family dinner table.

"The traditional isolated way that many schools have functioned," argued Mavis G. Sanders, "is anachronistic in a time of changing family demographics, an increasingly demanding workplace, and growing student diversity" (Sanders 2005). More and more STEM educators are de-isolating schools for these changing times.

Festivals, events, showcases, family night, science fair, robotics competitions, and other means of dismantling the walls between schools and communities generate buy-in. One of the biggest and best family STEM events is Minnesota's STEM Day at the State Fair (*http://stemdaymn.org*). Annually since 2009, dozens of exhibitors and stage performances enrapt many thousands of families from across the North Star State, supported by select corporate sponsors. Their neighbors to the south found the concept so appealing that since 2012 the Iowa Governor's STEM Advisory Council has sponsored STEM at the State Fair as a means to access the broadest and most diverse an audience gathered in any one place throughout the state (*www.iowastem.gov/STEM-Day-Fair*). Exhibitors from industry, higher education, K–12, trade organizations, and more provide scintillating learning experiences to children and their families. Popular exhibits include the Pella Window company's baseball-proof glass door, Rockwell Collins' flight simulator, and the University of Iowa

Photos courtesy of the author.

STEM at the Iowa State Fair, a premiere community engagement event, draws thousands of visitors each year for hands-on activities and awareness.

Creating a STEM Culture for Teaching and Learning

Medical College's Laparoscopic Surgical Trainer. By sunset on the fair, some 10,000 children and adults take part in one or more exhibits. Surveyed adults rate STEM at the State Fair an average 8.9 out of 10 for excellence. Children exit "more interested" in STEM at about an 80% frequency, with over 90% rating STEM at the State Fair "a lot" of fun.

What works at the statewide level scores even bigger locally. Community STEM events abound, driven by schools as well as employers, economic developers, libraries, universities, zoos, museums, churches, and combinations thereof. iExploreSTEM (*http://iexplorestem.org*) was a festival concept hatched in 2011 that has spawned replicas throughout Iowa and across the Northwest United States. Evidence of effect is similar to the impact of STEM at State Fair: Adults rate the value of iExploreSTEM at 8.7 on a 10-point scale, and 82% of children rate it "a lot" of fun with 79% expressing more interest in STEM as a result.

Designers of iExploreSTEM developed a detailed manual for community STEM festivals that includes a timeline, essential partners, funding sources, form templates, evaluative considerations, and more at *http://bit.ly/2oENyaV*. Plenty of other resources exist for anyone contemplating a bridge event that connects schools with communities in the name of STEM. For example the American Association of University Women (AAUW) coaches local chapters on how to host a STEM event (see *www.aauw.org/resource/workshop-hosting-a-stem-event*). North Carolina's STEM Community Collaborative released a 2012 Guide to STEM Community Engagement that features a systematic "community visioning and design process"(*www.ncpublicschools.org/docs/stem/resources/diy-guide.pdf*). Finally, the Science Festival Alliance is a membership organization of planners and executors of community science festivals. Among its resources is a Neighborhood Science Toolkit full of examples, templates, activity ideas, and even signage samples, which can be found at *http://sciencefestivals.org*.

A Community STEM Event Par Excellence

The first USA Science and Engineering Festival took place in Washington, D.C., in April 2010, and Dr. Gina Schatteman, on leave from her faculty post in Physiology at the University of Iowa for a Fellowship at the American Association for the Advancement of Science, helped plan it. "We regularly celebrate sports and the arts in this country, but rarely STEM," Schatteman explained, "It seemed overdue that STEM be included in our cultural landscape."[1] The USA STEM festival had planted a seed in Schatteman's mind about scaling such an event to the local level. Upon return to her home state, Schatteman discovered a convergence of interest along the same line of thinking from the State Hygienic Laboratory and an inter-university Iowa Math and Science Education Partnership (IMSEP). The three parties brainstormed a homegrown, scalable version of the national festival and iExploreSTEM was born.

The first iExploreSTEM festival was held on the lawn of the State Hygienic Laboratory in Coralville, Iowa, in September 2011. It featured 37 tent-based activities, three mobile

---

1 Quotes in this section are from the author's personal communications with Schatteman on June 2016.

labs, an "elevator pitch" competition, two interactive stage presentations, and a tour of the State Hygienic Laboratory. Exhibitors included representatives from universities, corporations, nonprofit organizations, schools, museums, and the media. Up to 1,000 visitors were expected, though Schatteman admits, "I was terrified that no one would come." In fact nearly 2,000 visitors showed up despite inclement weather, giving the team its first inkling of the thirst for such events in communities across the state and beyond.

The design and intent of iExploreSTEM was and continues to be about exposing children and their families to the STEM educational pathways in their midst, and to the careers right in their own backyards that await. Thus it is a very personal, locally relevant bridge-building experience both for planners as well as participants. Instrumental to the event's success is community ownership—planning from afar by detached organizers is a recipe for failure. And so is poor logistical planning. "The experience of neither the exhibitors nor attendees should be tainted by poor organization," cautions Schatteman. Much like a restaurant whose reputation is on the line with each dinner served, subsequent STEM events will depend on healthy relationships born of respect and organization in the early going. And, planners, exhibitors, and support crew all must commit to a free event in order to draw the most diverse and deserving of audiences.

Planners do well to incorporate an assessment plan into developments right from the start. Not only do exhibitors and sponsors rely on evidence to provide return on investment, but future events evolve under the bright light of measurement and feedback. The first iExploreSTEM was assessed by a university team from the IMSEP, and since then Schatteman has engaged a National Science Foundation consortium of science festival evaluators, EvalFest (*www.evalfest.org*), for protocols and instruments to gauge effect. She continues to value three primary metrics: enjoyment and learning on the part of children and families, exhibitor satisfaction, and achievement of event goals, including sheer numbers of attendees as well as their breakdown by gender, ethnicity, race, demography, and geography. One more measure that Schatteman considers is sustainability, stating that "If I am not around next year, will the community organize the festival again?"

There were 44 community STEM festivals enlightening over 13,000 citizens across Iowa in 2016, all built on the iExploreSTEM model. Schatteman has expanded iExploreSTEM throughout the Northwest United States. Her advice for others contemplating a STEM festival? "Start early," and "Involve as many partners as you can. Many hands make light work and you maximize community buy-in." Naturally her coaching tips include taking advantage of the planning guides available through iExploreSTEM or the Science Festival Alliance (*http://sciencefestivals.org*) and elsewhere. Importantly, events need to feel local and fresh, by ensuring that exhibitors reflect the demographics of the community and understand the nature and mission of the day (hands-on fun versus static displays, for example). And don't overlook the significance of site and date. "Avoid holding your festival in a school building if at all possible" because in Schatteman's experience, "No one wants to go back to school on Saturday." Nor "will anyone come if it's the day of the Iowa vs. Iowa State football game."

An eco-lodge design for Marrakesh, Morocco, by Aziza Chaouni is shown. As do architects of STEM partnerships, Chaouni brings a diverse voice to the table.

Architects of STEM

The inspiring architect Aziza Chaouni devotes her life's work to reconnecting human environments to the landscape—relinquishing a bit of control back to nature. She pulls together eclectic teams of local communities, experts in geology, hydrology, botany, structural engineering, and landscaping, to build restorative, harmonious habitats. Aziza's work parallels the modern STEM builder committed to reconnecting schools and their communities after a knowledge economy-driven fissure abetted by school buses. STEM's counterparts to Aziza dwell in that opportunity gap between the arithmetic rate of change in schools and the logarithmic pace of business. They, too, enlist diverse arrays of experts in unifying the worlds of education and industry as architects of partnerships, messaging, and community events. They do for schools what all-day breakfast did for McDonalds—strengthen the brand by adapting to an ever-evolving environment. STEM, all day.

References

Drucker, P. 1992. *The age of discontinuity.* Piscataway, NJ: Transaction Publishers.

Heiden, E. O., M. Kemis, M. Whittaker, K. H. Park, and M. E. Losch. 2016. Iowa STEM monitoring project: 2015–2016 annual report. Cedar Falls, IA: University of Northern Iowa,

Center for Social and Behavioral Research. Modi, T. 2011. *Living in the innovation age: Five principles for prospering in this new era.* Springfield, VA: TekNirvana.

Modi, T. 2011. *Living in the innovation age: Five principles for prospering in this new era.* Ahmedabad, India: TekNirvana.

National Academy of Engineering Committee on Public Understanding of Engineering Messages. 2008. *Changing the conversation: Messages for improving public understanding of engineering.* Washington, DC: National Academies Press.

Sanders, M. G. 2005. *Building school-community partnerships: Collaboration for student success.* New York: Skyhorse Publishing.

Spaulding, M., M. A. O'Leary, and J. Gatesy. 2009. Relationships of Cetacea (Artiodactyla) among mammals: Increased taxon sampling alters interpretations of key fossils and character evolution. *PLOS ONE* 4 (9): e7062. doi:10.1371/journal.pone.0007062.

Yoder, B. L. 2015. Engineering by the numbers. Working Paper, American Society for Engineering Education, Washington, DC. *www.asee.org/papers-and-publications/publications/college-profiles/15EngineeringbytheNumbersPart1.pdf.*

CHAPTER 4

School–Business STEM Partnerships

A signature of the STEM awakening has become the school–business connection. What was once the sole domain of career-technical education, workplace applications of content and on-the-job learning permeate the core disciplines of math and science in a STEM school culture. This chapter characterizes some thriving models that bridge the widening education–industry chasm, defining the roles of teachers, students, and business partners. The emergence of work-based learning with its potency in soft skills (also known as employability skills, success skills, and human skills) development holds great promise.

Brewing Up a Partnership

For six weeks in the summer of 2015, biology teacher Ehren Whigham externed at the Confluence Brewing Company in Des Moines, Iowa, working with its microbiologist to monitor yeast health while screening for bacterial contamination. An outcome of his summer in industry was to find new ways to use yeast back at school to teach concepts like cell communication, population dynamics, evolution, and fuel ethanol production. But a funny

thing happened on the way back to his Roosevelt High School classroom—Whigham learned that Confluence wanted to branch into root beer and gifted him a brew kettle. His students transformed the lab into a distillery (nonalcoholic) and as part of their fermentation unit, brewed up various batches until they got it right. Today their partner Confluence cannot make the stuff fast enough—the first batch sold out in 36 hours, with some of the proceeds channeling back to Roosevelt for laboratory supplies. And they'll need the expanded capacity—students at Roosevelt (the Roughriders) are now working on recipes for Rider Cream Soda, Rider Cola, and other variants while sharpening their graphic design and marketing chops as young entrepreneurs.

Whigham's partnership with Confluence is a sign of the times. The walls around schools are coming down—figuratively in many places, literally in a few—to address an essential question of today's student: "When am I ever gonna need to know this?" We can tell them of course, and goodness knows that well-meaning teachers have striven to devise lessons and lab experiences linked to applications beyond classroom walls since John Dewey implored us to crack the isolation of schools (Chapter 1). But to engage with real, live workplaces is fail-safe, sure-fire, slam-dunk cinch that the content has meaning beyond the classroom. School–business partnerships are to education what antibiotics have meant to medical science, which was a revolutionary and almost entirely positive development. So much so that further study using experimental and control groups presents an ethical dilemma of jeopardy for the unfortunate individuals assigned to the control group. The remainder of this chapter examines when partnerships work and profiles common characteristics of the best partnerships, with examples.

Rider Root Beer is brewed, bottled, and marketed by a partnership between Confluence Brewery and the students at Roosevelt High School in Des Moines.

Reprinted with permission from KCCI Des Moines.

When School–Business Partnerships Work
Distributed Load and Shared Wealth

Partnership has become a term *du jour* these days, joined by the likes of *catalyst, crowd-source, chillax, photobomb, optics, upcycle, funemployment, yolo, unmeeting, framily, staycation, binge-watch, sea change,* and let us not forget *bromance.* This is a mere sampling of the trending vernacular of pop culture. Some have arisen of late to fill a void in our yet-to-catch-up English language. (A decade ago, there were no friends-and-family cell phone bundle discounts, thus no "framily" plan.) Others have taken on new meanings, such as *catalyst,* which we knew from Chem 101 as something that hastens the reaction of substances but has come to refer to something that moves people to action. But the word *partnership* has become ubiquitous across the landscape, so widespread in use as to now be prone to co-opted use kind of like STEM.

For example, consider the precarious partnerships between professional sports teams and their host cities. According to a 2015 story in the online journal *MarketWatch* titled "5 Cities Getting the Worst Deals From Sports Teams," 22 NFL teams have managed to extract hundreds of millions of dollars each from their "partner" cities over the last couple of decades. One of the worst deals, according to the article, is in Minneapolis where taxpayers ponied up half the $1 billion for a new football stadium, lest the team act on a veiled threat to relocate to sunny California. Today the Vikings are official partners of the Minneapolis Downtown Council, and although there is certain to be some urban revival thanks to the new stadium, by what measure is it a partnership (Notte 2015)?

The same might be asked of another high-profile partnership—that between Western ranchers and the Bureau of Land Management (BLM). The BLM–ranchers partnership allows cattle ranchers to graze their livestock on public property for a fee, so long as they abide by regulations such as the number of cows allowed in an area, how much they can graze, and how many cows can be clustered around a stream. But the partnership, more akin to a truce, boiled over in an armed standoff in Southeastern Oregon in 2016 (Burns and Schick 2016). Someone wasn't feeling much like a partner.

School–business STEM partnerships work when the load is distributed and the wealth is shared. Sherri Torjman of the Caledon Institute of Social Policy defined the essential characteristics of partnerships in her historic 1998 paper "Partnerships: The Good, The Bad and The Uncertain" in

School–business partnerships work best when both sides have an equal stake in the game.

this way: "There is no partnership without a sharing of risk, responsibility, accountability and benefits." The partnership described at this chapter's opening is a good test case for Torjman's r.r.a.b. definition.

Roosevelt High School's partnership with Confluence Brewing works because Ehren took a significant risk in crossing a border to immerse in the world of commerce one summer. Likewise, Confluence risked precious human and financial resources on Ehren, bringing him up to speed, making him useful. Roosevelt then took responsibility for broadening the curriculum to accommodate a new unit, while Confluence held up its end on the production side, agreeing to bottle the students' concoction. As to accountability, both parties are held to a highest of standards—product quality and marketability. And when it comes to sharing in benefits, dollars are flowing inward to both partners while Roosevelt students learn new STEM skills and Confluence gets a research and development department. Neither partner feels tethered to an unfair toll like Vikings fans might, nor bridled by burdensome rules as some Western ranchers perceive. That is what partnerships feel like. Later in this chapter, the Torjman r.r.a.b. litmus test will be applied to other exemplary school–business STEM partnerships for their value to both parties.

Evidence Informs Practice

Another way partnerships work is by accumulating evidence that feeds back to improving practices. Students who take part in long-term investigations of community issues with local business partners make significant gains in academic engagement, 21st-century skills, and STEM content according to an evaluation of a partnership called the STEMester of Service program of the Youth Service America (YSA). The YSA's Education Director Scott Ganske, in his essay "Students as Innovators in STEM" states that students who take up critical community problems through the application of learning in STEM content transform their research into ideas and actions (Ganske 2015). The finding aligns with and supports all that we know of learning theory as a lens not just for the design and critique of school–business partnerships but for STEM at large, as addressed in the Learning Theory subsection of Chapter 1 of this book, where the authors of *How People Learn* posit that "the ultimate goal of schooling is to help students transfer what they have learned in school to everyday settings of home, community, and workplace."

The accumulated evidence for bolstering and expanding partnerships is so strong that the National Academies of Science, Engineering, and Medicine, through a 2016 study conducted by the Committee on Improving Higher Education's Responsiveness to STEM Workforce Needs, issued something of a declaration: "Systemic connectivity needs to include more structured industry participation in students' educational experiences," and there is a need "for greater opportunity for student internships and apprenticeships throughout the STEM-related workforce" (National Academies of Sciences, Engineering, and Medicine 2016, p. 37). Because not only do learners progress in content knowledge and skill sets, they also pick up a number of associated benefits, captured nicely by an Australian Council for

Educational Research 2011 report "The Benefits of School–Business Relationships." Some benefits are as follows:

- Improved student relationships with peers and family
- Greater self-esteem, confidence, and self-awareness
- Higher aspirations for the future
- Improved goal setting, teamwork, and conflict resolution skills
- Enhanced leadership skills
- Greater ability to learn independently
- Healthier lifestyle habits
- Greater respect for past generations
- A more positive outlook on life
- Increased awareness of the work of community groups

A note of caution is in order amid all of the celebration of partnerships. Like antibiotics (mentioned earlier as a revolutionary and almost entirely good development), partnerships between schools and businesses can be overprescribed. Educators do well to enter into partnerships with eyes wide open, as advised by Alfie Kohn in his timeless 2002 piece for *Phi Delta Kappan* titled "The 500-Pound Gorilla," "When corporations can influence the nature of curriculum and the philosophy of education, then they have succeeded in doing something more profound, and possibly more enduring, than merely improving their results on this quarter's balance sheet. That can happen when businesses succeed in creating 'school-to-work' programs" (Kohn 2002). Kohn's point is that business motivations for engaging in education can be, obviously, very different from educators' motivations for engaging with business. Part of everyone's responsibility is to understand each other's motivations.

Standards Support School–Business STEM Partnerships

A third powerful impetus and guidepost for school–business STEM partnerships is standards. Especially the practices of design, problem solving, critical thinking, communication, and others that transcend mathematics, science, and technology standards. Often by teachers' own admissions, the skills of the discipline can get short shrift in a crowded curriculum, in favor of content coverage, though the employment sector would likely flip these priorities. But the journaled confessions of teachers spending their summers in workplace externships across Iowa reveal a course adjustment to come back at the classroom. "The managers are looking for people that will go and find out how to solve problems without relying on everyone else," observed a mathematics teacher three weeks into work at a tractor manufacturer, resolving, "One thing that I will do WAY more next year because of this is make sure we do problems that don't provide all the information they need to answer it. This is so much

more 'real world' than the math problems that students have gotten used to!"[1] For a science teacher working at an aerospace electronics manufacturer, clarification came in the form of an assigned complex coding task to help departments process maintenance information. "I am going to admit it, employability skills (in my classroom) are not always the highest priority. I feel a lot of pressure (some perceived, some real) to make sure that I make it all the way through my assigned curriculum," she admits, only to discover on-the-job, "The technicians need to be able to easily input the data, the facilities and supervising engineer need to be able to read the data, and it all needs to happen automatically. From initial conception, to rough draft, to beta testing, everyone needs to be (and should be!) happy with how things are working." The result is a shift heading back to school. The teacher explains, "I am going to try and be more aware of introducing 'Stakeholders' in the student's projects that all have a need for the information and all bring a constraint to the party."

Standards support these teachers' workplace epiphanies. The *Next Generation Science Standards* provide quite a lift when it comes to partnering with business. Three dimensions encircle educators and their students in a comprehensive and integrated approach to science

and engineering teaching and learning (NGSS Lead States 2013). First, *science and engineering practices* are the combined skills and knowledge that scientists employ as they investigate and build models and theories about the world and a key set of engineering practices that engineers use as they design and build systems. Where better for students to witness and acquire such practices than the workplace? Second, *crosscutting concepts* emphasize the need to consider not only disciplinary content but also the ideas and practices that cut across the science disciplines. On the job at a manufacturing floor or in the accounting suite no bells separate math from science or English; rather, job sites are inherently interdisciplinary. And third, *disciplinary core ideas* invite teachers to relate to the interests and life experiences of students or be connected to societal or personal concerns that require scientific or technological knowledge. And the reservoir for drawing such relatable lessons pours from workplaces that surround any school.

Down the hall, rest assured that math teachers and technology teachers, too, have the backing of their universally embraced standards for engaging in partnerships. The *Common Core State Standards for Mathematics* set targets for eight mathematical practices that are

---

1 Quotes in this section are from the author's final report to the National Science Foundation's Innovative Technology Experiences for Students and Teachers program (grant number DRL-1031784). The report is titled "Real World Externships for Teachers of Mathematics, Science and Technology, 2012 to 2016."

best taught in the context of workplace applications (NGAC and CCSSO 2010). Consider how readily achievable the standard "Apply mathematics to solve problems arising in everyday life, society, and the workplace" for a teacher familiar with workplace applications of geometry, algebra, and calculus (NGAC and CCSSO 2010). And imagine the power of relevance to learners asked to statistically prove their hypotheses about menu preferences at a local restaurant by time of day or year to fulfill the standard "Compare the effectiveness of two plausible arguments, distinguish correct logic or reasoning from that which is flawed, and—if there is a flaw in an argument—explain what it is" (NGAC and CCSSO 2010).

The technology standards of both the ITEEA and ISTE mentioned in Chapter 1 guide technology educators to "evaluate and select information sources and digital tools based on the appropriateness to different tasks" (ISTE Student Standard 3.C; ISTE 2016). They also state that "students will develop an understanding of Design. This includes knowing about the role of troubleshooting, research and development, invention and innovation, and experimentation in problem solving" (ITEEA Standards for Technology Literacy 10; ITEEA 2007). By their very nature, technology standards tend to and rely on connecting classroom experiences to the world of business.

Types of School–Business Partnerships

Models bloom across the country, both major corporate as well as local "mom and pop" school–business connections. Resources and ideas are plentiful, but *caveat emptor*—their evaluative evidence, applicability or generalizability, and motivations all need to be carefully considered.

Guidelines for Partnering

The U.S. Chamber of Commerce has been busy in this space for a decade, producing first a collection of major corporate examples of partnerships in the 2005 report *Business-Education Partnerships in the United States: Committed to the Future*. The report details how companies like Marriott, IBM, Oracle, and others on that scale provide STEM education resources at the local and systems level. Transitions to Teaching is IBM's program to retrain employees to enter the teaching profession. Oracle trains and supports teachers to deliver courses on database and Java programming. And Marriott partners with DECA (Distributive Education Clubs of America) to provide internships, scholarships, and conferences. Then in 2011, the U.S. Chamber produced the report, *Partnership Is a Two-Way Street: What It Takes for Business to Help Drive School Reform*, which highlights three cases: Nashville, Tennessee; Austin, Texas; and the Massachusetts Business Alliance for Education (MBAE). The education and business sectors in Nashville have come together to offer *Smaller Learning Communities* grants to support high school redesign. In Austin, school and business leaders formed an alliance to build strategies (such as Saturday family Financial Aid seminars) and measures (graduation and college-going rates) to boost postsecondary transitions. And the MBAE lobbied for the eventual adoption of Common

Core in Massachusetts (Hess and Downs 2011). Finally the U.S. Chamber rounded out these guides with the 2016 *The Path Forward: Forging Partnerships to Improve Education*, which coaches would-be individuals and organizations step-wise toward partnering that align fairly well with Sherri Torjman's risk, responsibility, accountability, and benefits (U.S. Chamber of Commerce 2016).

Of a less commercial slant, the U.S. Department of Education announced a High School Redesign initiative in 2013 intended to challenge high schools and their partners to rethink teaching and learning, putting in place learning models that are rigorous, relevant, and better focused on real-world experiences. Beacons of the vision cited by the department include the Pathways in Technology Early College High School (P-TECH), begun in Brooklyn, New York, where students graduate from high school with an associate's degree in technology; and the High Tech High network begun in San Diego, California, known for integrated, project-based learning and required internships in the community. Another recent initiative of the U.S. Department of Education around STEM includes a grant program for STEM Innovation Networks to recruit and prepare STEM teachers for a mission of college-and-career ready graduates. The Tennessee STEM Innovation Network is a fine example, launching "STEM Platform Schools" to showcase best practices in STEM education, and Regional STEM Innovation Hubs that connect local schools to informal education resources, businesses, higher education, and others to amplify local innovations.

An abundance of guidelines, toolkits, roadmaps, and strategic plans have come forth from state and local education organizations, many times reinventing the wheel in coaching school–business partnerships. Here is a sampling of some of the better wheel reinventions because of their stepwise recommendations, examples, and evaluative tips:

- Montana's Department of Public Instruction joined with State Farm Insurance to produce the comprehensive *Graduation Matters Montana: School-Business Partnership Toolkit* (available at *http://opi.mt.gov/pdf/GradMatters/12GMM_Business_Toolkit.pdf*).

- The school district of Anchorage, Alaska, alongside the Anchorage Chamber of Commerce created the dated but specific *Handbook for Developing and Maintaining Successful Partnerships* (available at *www.asdk12.org/forms/uploads/sbp_handbook.pdf*).

- The North Platte, Nebraska, school district developed the practical North Platte School District School-Business Partnership Program (learn more at *http://bit.ly/2pqkGDp*).

- The Winston-Salem/Forsyth County Schools of North Carolina, in partnership with the Winston-Salem Chamber of Commerce, generated the detailed *Partnerships in Education: A Guide to Establishing, Building, and Maintaining Partnerships That Last* (available at *www.winstonsalem.com/wp-content/uploads/2013/07/partnership-how-to-guide.pdf*).

- The Iowa Governor's STEM Advisory Council, in response to community clamor for guidance on establishing and nurturing partnerships, wrote the state-specific *Building*

Connections Between Education and Business, which others may find useful (available at *http://bit.ly/2oghGsh*).

Examples of Partnerships

STEM BEST (Business Engaging Students and Teachers) originated in 2014. A program of the Iowa Governor's STEM Advisory Council, the goal of BEST is to unite the expertise of public and private sectors to strengthen the continuum from school to careers. It involves a framework built on three key attributes: (1) education driven by industry need; (2) rigorous, relevant, and dynamic STEM curriculum; and (3) authentic partnerships. Singular school–business arrangements as well as multipartner consortia have developed BEST partnerships. Each one involves some variation of the timely and burgeoning work-based learning experience. The Rocket Manufacturing local network example of Chapter 2 is a single school model of a STEM BEST, where students at Rock Valley High School fabricate custom-ordered metal work for area manufacturers. The Roosevelt High School Confluence Brewery profile that launched this chapter would be a supreme BEST model, though it congealed without aid. A rural

STEM BEST Students at Northeast Goose Lake High School collaborate on architectural design with a local contractor as a STEM BEST partnership in Iowa.

schools consortium of four high schools and seven industry partners make up a Northeast Iowa BEST model that immerses students in half-day professional-based learning experiences custom-fitted to their interests. Through a $25,000 grant (cost-matched by applicants), STEM BEST partners develop STEM curriculum, cultivate community partnerships, professionally develop their teachers and partners, create sustainability plans, craft financial structures, and serve as models in disseminating their work to others. Iowa's STEM BEST owes a nod to the aforementioned P-TECH model as well as to the Blue Valley, Kansas, CAPS (Centers for Advanced Professional Studies) model for inspiration in the design. (More information about STEM BEST may be accessed at *www.iowastem.gov/STEMBEST*.)

STEM Teacher Externships have been operating across Iowa since 2009. Ehren Whigham of Rider Root Beer fame entered the world of partnerships via his own

externship, as so many do. The program operates under the governor's STEM Advisory Council in answer to that age-old question of students everywhere, "When am I ever gonna need to know this?" The inner workings of the externships program are discussed Chapter 8 (STEM professional development) but now is an appropriate place to highlight the partnership aspect. Businesses and other workplaces opt in for an extern for both altruistic school-improvement reasons as well as self-serving motives, as captured recently by the president and CEO of Kemin Industries, Dr. Chris Nelson, "Kemin Industries has taken in Externs since the program's launch in 2009 for two reasons. First, teachers make real and significant contributions to our operations in the lab, out in the field, and on our teams. And second, even more importantly we know that teachers take back to school innovative teaching ideas that improve learning experiences that in turn produce career-ready graduates for Kemin and other employers across Iowa."[2] A teacher who worked with the Department of Natural Resources through the summer extended the partnership to habitat restoration around the school grounds over sub-

Extern Laura Condon, math teacher at Estherville Lincoln Central high school, spent the summer at GKN Wheel analyzing production-line efficiency and safety.

Photo courtesy of the author.

sequent years. A plastics manufacturer enlisted an extern and his students back at school to design and test models for improving the switch-out of pigment injectors, a costly downtime necessity at the plant. One of the outcomes monitored closely by program evaluators is the sustained school–business partnerships that ensue from a summer externship. Of the several hundred teaches who have participated to date, about three-fourths maintain connections to their workplace hosts in some fashion. (For additional information about externships, see *www.iowastem.gov/externships*.)

Curricular Programs define the entry-point for businesses to partner with schools. Iowa's Scale-Up initiative of the Governor's STEM Advisory Council supports the competitive selection and subsequent implementation of exemplary STEM programs that, among a variety of criteria, prove adept at connecting educators and industry partners. Dozens of programs, ranging from computer science clubs to elementary school engineering, and from robotics to the science of racing and more, each create prescribed space for uniting students and their teachers with community professionals. Project Lead The Way,

---

2 Quotes from Nelson in this section are from his remarks at the Kemin Outstanding STEM Teacher award ceremony on June 28, 2016, in Des Moines, Iowa.

for example, assigns recipients the happy task of assembling a program advisory board comprising area engineers and technicians. FIRST (For Inspiration and Recognition of Science and Technology) paves the way for expert mentors to serve as coaches and advisers for youth robotics teams. HyperStream asks school computer science clubs to partner with information technology executives in the vicinity. And the National STEM League's Ten80 Racing Challenge provides a scaffold for learners to connect to professional technicians for advice and consult. Iowa's Scale-Up program, begun in 2012, reaches thousands of educators

A Pint-Size Science participant works on his project. Pint-Size Science is one of Iowa's Scale-Up programs bringing top STEM learning experiences to students across the state.

and tens of thousands of youth each year. (For additional information about any of these programs or the initiative at large, see *www.iowastem.gov/Scale-Up.*)

STEM Teaching Award, a classic engagement point for business, conveys appreciation (and some believe incentive) for high-quality education while gaining positive public relations. A great many businesses and organizations across the United States sponsor teaching awards in the disciplines of mathematics, science, engineering, and technology, some local, others state and national. The I.O.W.A. (Innovative, Outstanding, Worldly, Academic)

I.O.W.A. STEM Teaching Award recipient Dirk Homewood (second from left), engineering teacher at Cedar Falls High School, receives the award from Kemin Industries President and CEO Dr. Chris Nelson (far left), Governor Terry Branstad (second from right) and Governor Kim Reynolds (far right).

A view of STEM Day at the State Capitol. The event has become an annual rite in Iowa.

STEM Teaching Award is a business partnership with the Governor's STEM Advisory Council by which Kemin Industries sponsors the promotion, selection, award, celebration, and recognition of six outstanding educators each year. A brainstorm and personal commitment of Nelson, the award "is just one of the ways we enjoy honoring the hard work and dedication of exceptional teachers in our state," Nelson explains. "These teachers deserve recognition for preparing today's students to become tomorrow's workforce, which includes an abundance of STEM opportunities."

A final example, **events** are a frequent platform for school–business partnerships in STEM for the networking payoff they bring. STEM at the Capitol, STEM at the Gym, STEM at the Fair, and STEM Film Festivals are all successful partnership events. The previous chapter mentioned festivals such as the Minnesota State Fair, the USA Science and Engineering Festival, and the iExploreSTEM community events. What they all have in common are numerous corporate and other nonschool partners contributing to a school mission, which is to educate the next generation. One of the many reasons they are successful is turf—often a neutral nonindustry, non-education venue owned by neither entity, but by the people they serve. Everyone is equal and welcome at a STEM event.

Partnership Profiles

To cap off this chapter on school–business STEM partnerships, a series of professional partners have willingly been profiled, each advancing STEM education by aggressively

uniting the worlds of business and education by demonstrating the roles, responsibilities, accountability, and benefits that make a partnership sing. They are sequenced from the school level, to regional, to state, then global.

Local-Level School–Business STEM Partnerships

How many schools can you name that build in professional development time for instructors to meet with leaders in business and industry so they can learn from and partner with them on student learning projects? It's standard fair at Central Campus in Des Moines, a regional academy serving Central Iowa. The school's director, Aiddy Phomvisay, expects every teacher to seek out and nurture partnerships in order to provide a relevant and rigorous learning experience that responds to community needs. Established partners of the school include national names such as Cargill, John Deere, Cisco, DuPont Pioneer, Principal Financial, and IBM, as well as regional companies including Unity Pointe Health, Mid-American Energy, and Lumberman's Supply. The last two partners along with Hubbell Homes and others have formed a *skilled trades alliance* to fundraise for equipment, technology, and internships. Phomvisay expects teachers to learn from business partners to update curriculum, student projects, and assessments. Students are provided with as many authentic learning experiences as possible. "The ideal situation," said the director, "is for our students to be in the field with these businesses on a regular and ongoing basis." It's a golden opportunity for the region's private sector, too, he believes—"Investing in these students and programs are value added propositions that strengthen business and economic and community betterment" (personal communication).

Regional School–Business STEM Partnerships

Business and trade organizations are beginning to see local schools as primary partners in workforce development. That is certainly true for the Opportunity2 economic development group in south central Iowa whose leaders resolved to counter the brain-drain from their region by equipping teachers to sell the area's great career opportunities and to become strong advocates for employers. In partnership with Des Moines Area Community College (DMACC) Business Resources, they launched a summer professional development program for teachers called Teaching for the Workplace, consisting of industry tours, briefings by human resource managers as well as plant managers, and insights on the application and interview process so as to help students transition from the academic world to the workplace. All business sectors of the region participate, including advanced manufacturing, health care, IT, and finance, plus rural electric cooperatives. The goal of the program, according to DMACC Business Resources executive director Kim Didier, is to give the teachers the broadest picture of the business landscape in the region and the various pathways for students of all skills and abilities. "Many of the businesses hosted the

group for one day and sponsored lunch," Didier said of the workshop, "They also provide tours and insights into what they are looking for from prospective employees."[3]

It was an easy sell to regional businesses, according to Didier, "The businesses were eager to support the initiative as they saw a direct correlation to their need to recruit employees." Selling the concept to teachers involved building up a collaborative environment through informational meetings along with a battery of promotional tactics, including flyers to the schools, social media posts, newspaper articles, and contacting principals. As a result, the profile of participating teachers is wide, ranging from elementary to middle, and high school teachers of math, science, industrial technology, family and consumer science, special education, and other subjects.

"This out-of-the-classroom experience," surmised Didier, "provides educators with the tools to assure that essential subject matter is being taught and essential knowledge and skills are being learned to meet the needs of today's workforce."

State-Level School–Business STEM Partnerships

One of the leading statewide STEM organizations when it comes to engaging business at the partner level is Washington STEM. Founded in 2011, Washington STEM was conceived of by the Washington Roundtable, an association of the state's largest employers. "The Boeing Company, Microsoft, and McKinstry in particular were deeply involved in early (and current) leadership and financial backing of Washington STEM," recalled Caroline King, CEO and Strategy Officer for Washington STEM, "and members from all of these companies remain on our Board of Directors, in addition to members from Saltchuk, Lease Crutcher Lewis, Alaska Airlines, and Accenture."[4]

The commitment of the state's private sector permeates Washington's STEM organization laterally to the borders and vertically into communities. Ten regional STEM Network Leadership teams canvas the state, each populated with local and regional business representatives. At the program level, volunteer expertise and corporate sponsorships make significant impact. "We launched a new Engineering Fellows program to pair fifth-grade teachers with engineers as they design real-world challenges for students next school year," King shared, adding, "and Microsoft staff is supporting us during a hackathon to increase resources for teachers

---

3 Quotes in this section are from the author's personal communications with Didier in June 2016.
4 Quotes in this section are from the author's personal communications with King in June 2016.

navigating career connected learning." Regional employers such as Avista in Spokane, the Puget Sound Naval Shipyard in Bremerton, and Stemilt Growers in Wenatchee all partner in Washington STEM's mission to inspire future STEM talent across the state. Partners also contribute to Washington STEM by sponsoring the state's annual STEM Summit, advocating for policy change in the state legislature, launching a new investment fund focused on girls and STEM, and partnering to produce a STEM community event.

What their business partners share is a sense of corporate citizenship—to be part of the solution in creating pathways for students to be job-ready for the STEM careers that await across Washington (the state is ranked second in creation of STEM jobs according to King). A short-term benefit for employers is the technical assist that many students can bring through internships, while a long-term benefit is building a pipeline of inspired and prepared talent to meet the growing demand for workers.

Keeping Washington STEM's business partners connected and committed involved providing evidence of return on investment as well. Metrics of high value include the following:

- Increase (and diversify) students earning postsecondary/technical STEM degrees.

- Increase (and diversity) students inspired in STEM by middle school.

- Increase the number of women and people of color who have the interest and preparation to pursue STEM pathways and careers.

Results ensure a strong future of business-education STEM partnerships in Washington. In fact, King considers business the "third leg of the table along with education and community" in cross-sector linkages that provide a scaffold for expanding a career connected learning program. Evidence begets even more engagement on specific and scalable efforts with business partners, "such as our Engineering Fellows program or business–education partnership tools such as Wenatchee Learns," said King, "that support the entrance of Washington students into STEM careers."

The CEO's advice for other states or regions that seek to establish STEM business partnerships? "Assess the motivations of business clearly and develop some concrete ways to partner—both short and long term," coaches King. "Most businesses have an interest in a strong, sustainable workforce and in corporate citizenship. Show how your efforts will address their needs… and provide concrete ways for businesses to lead and help deliver the solutions and own the success." And finally, "Celebrate your business partners through media and other recognitions in the community." These are verdant insights from the Evergreen State.

National and Global School–Business STEM Partnerships

A handful of U.S. multinational corporations have strongly embraced and supported the STEM education movement, and one of the best among them is Rockwell Collins (RC).

Rockwell Collins' STEM Day brings Rockwell Collins' product demos, facility lab tours and STEM activities to youth in our community. Employee Heidi Kiser helps a child learn to code through one of the several hands-on activities offered during the event.

Its business–education partnership program recently celebrated its 25th anniversary. From humble beginnings focused exclusively on the home school districts of world headquarters in the Cedar Rapids/Marion area of Iowa, the company now invests an estimated $3 million per year globally in STEM programming alongside nearly unquantifiable human resource investment in the form of employee-volunteers and executive leadership.

Rockwell Collins' motivation for involvement in STEM education is simple: As the company grows, its need for high-quality job candidates expands as well. "Engineers are the life-blood of our company," explained Cindy Dietz, director of external relations, "and so building a pipeline is essential to developing a talent pool."[5] In the early 2000s, Rockwell Collins started to formalize STEM outreach and include other Rockwell Collins sites. By 2007, it had a formalized STEM strategy, dozens of educational partners, and multiple locations with employees volunteering to get students interested in STEM careers. They established guiding principles for choosing what programs to support based on alignment with a strategic vision, mission, and goals that constitute the company's formalized strategy. A priority is placed on programs or partners that emphasize engagement of students that are female, minorities, low-income, or high risk. The company also prefers to support programs already involved with partners in other areas of the company, for example an association with the United Way or colleges in which they have a relationship. Other favored program characteristics include the following:

• Focuses on the K–12 education arena

---

5 Quotes in this section are from the author's personal communications with Dietz in June 2016.

- Has an international or national scope, but with a local focus

- Offers programming that is hands-on

- Encourages teamwork and inclusiveness

- Engages the Rockwell Collins employee base meaningfully and successfully

Programs that pass muster with Rockwell Collins, earning financial support, volunteer support, or both are as follows:

- Strategic partner with FIRST

- DiscoverE

- Project Lead The Way (PLTW)

- Internships and job shadows

- Future City Competition

- STEM career fairs and presentations

Evidence of effect weighs mightily in sustaining that support. Rockwell Collins values evidence of engagement, retention, and success of minority, female, low-income, or high-risk students. They also place a premium on employee engagement, investment, and retention in local communities as a powerful side-effect of getting students interested in STEM careers. In fact the company has found, according to Dietz, that employees who get involved with STEM volunteering have a boost in morale and a renewed appreciation for their work at Rockwell Collins. Dietz and her company's President and CEO Kelly Ortberg lead by example—both volunteer their time and wisdom to the Iowa Governor's STEM Advisory Council.

Hard-earned insights for other global businesses inclined to invest in STEM education are readily volunteered by Dietz: "Establish a vision, mission, and guidelines with input from different areas of your company to articulate what is important." The foundation enables a critical next step to define the program goals and outcomes. "Share the plan in order to build acceptance and buy-in from all parts of the company," she advises. And now that many companies are decades into refining their STEM programs, Dietz cautions, "Do not reinvent the wheel; there are several programs and partners that are more efficient, have a solid focus, and have an established history and curriculum on what you might be trying to create."

Blurring Boundaries Between School and Business

"Is this Heaven?" asked Chicago Black Sox outfielder Shoeless Joe Jackson, played by Ray Liotta in the 1989 movie *Field of Dreams*. "No" answered Kevin Costner's character Roy Kinsella, "It's Iowa." Imagine the STEM community so effective at school–business

partnerships that boundaries blur, and students are overheard asking, "Is this business?" to answers of "No, it's school." The two places in a learner's life would be indistinguishable, as is the case in the fermentation lab at Roosevelt High School, or at a STEM BEST site, or back at school after an externship, or coding side by side with an IT mentor. With the backing of research, standards, guidelines, and every imaginable constituency, partnerships are chock-full of potential good, for students. As for business, John Martin, the president of Confluence Brewing, sums up his side of the equation, "Hopefully we are seen as a responsible and productive part of the community that is willing to give back" and while they're at it, partnering with school is "a good way to let people know who we are and hopefully let our values shine through in the process" (personal communication, June 2016). Cheers to school–business STEM partnerships.

References

Australian Council for Educational Research. 2011. The benefits of school-business relationships. *https://docs.education.gov.au/system/files/doc/other/benefits_sbr_acer_report.pdf.*

Burns, J., and T. Schick. *PBS Newshour.* 2016. Before Oregon's Armed Takeover, a Long-Brewing Dispute Over Rangeland Health. January 7. *www.pbs.org/newshour/updates/what-is-environmental-health-and-why-did-it-trigger-oregons-armed-takeover.*

Ganske, S. 2015. Students as innovators in STEM. In *Advancing a jobs-driven economy: Higher education and business partnerships lead the way*, ed. STEMconnector, 142–145. New York: Morgan James.

Hess, F. M., and W. Downs. 2011. Partnership is a two-way street: What it takes for business to help drive school reform. Institute for a Competitive Workforce. *https://www.*

uschamberfoundation.org/publication/partnership-two-way-street-what-it-takes-business-help-drive-school-reform.

International Society for Technology Education (ISTE). 2016. ISTE standards for students. *www.iste.org/standards/standards/for-students-2016*.

International Technology Education Association (ITEEA). 2007. *Standards for technological literacy: Content for the study of technology.* 3rd ed. Reston, VA: ITEEA.

Kohn, A. *Phi Delta Kappan.* 2002. The 500-Pound Gorilla. October. *www.alfiekohn.org/article/500-pound-gorilla*.

National Academies of Sciences, Engineering, and Medicine. 2016. *Promising practices for strengthening the regional stem workforce development ecosystem.* Washington, DC: National Academies Press.

National Governors Association Center for Best Practices and Council of Chief State School Officers (NGAC and CCSSO). 2010. *Common core state standards.* Washington, DC: NGAC and CCSSO.

NGSS Lead States. 2013. *Next Generation Science Standards: For states, by states.* Washington, DC: National Academies Press. *www.nextgenscience.org/next-generation-science-standards*.

Notte, J. *MarketWatch.* 2015. Opinion: 5 Cities Getting the Worst Deals From Sports Teams. July 12. *www.marketwatch.com/story/5-cities-getting-the-worst-deals-from-sports-teams-2015-07-17*.

Torjman, S. 1998. Partnerships: The good, the bad and the uncertain. Toronto: Caledon Institute of Social Policy. *www.caledoninst.org/Publications/PDF/partners.pdf*.

U.S. Chamber of Commerce, Business Civic Leadership Center. 2005. *Business-education partnerships in the United States: Committed to the future.* Washington, DC: U.S. Chamber of Commerce. *www.uschamberfoundation.org/sites/default/files/publication/ccc/BENcasestudy.pdf*.

U.S. Chamber of Commerce. 2016. The Path forward: Forging partnerships to improve education. *www.uschamberfoundation.org/sites/default/files/The_Path_Forward_Forging_Partnerships.pdf*.

CHAPTER 5

STEM Teachers, STEM Classrooms, and STEM Schools

Over the course of the early 21st century, STEM has enjoyed a bandwagon effect as the brand strengthens. Purists and pioneers have put structures in place to establish a certain standard that, in this chapter, upholds important labels such as STEM teacher (both in and out of school), STEM classroom, and STEM school. What brands a teacher or a classroom or a school as "STEM"? That is the value of standardized rubrics and systems of recognition to discern and honor best practices in preparing educators capable of conducting student-centered, active, problem-solving instruction making purposeful connections to local business partners. These structures, standards, and their effect are explored.

CHAPTER 5

STEM Bandwagon

STEM leads a conga line. Up until just a few years ago it was a lonely dance to a different acronym. But flip SMET to STEM and broaden the stakeholder community and voilá—an undulating spine of supporters have joined the parade. Today, STEM enjoys a bandwagon effect. Scholars of bandwagon-ology have found the bandwagon effect to result from the snowball effect, that is, winners accumulate followers the more they win. In their landmark 1987 paper "The Emergence of Bandwagon Effects: A Theory" in *The Sociological Quarterly*, Richard Henshel and William Johnston point out the influence of favorable pre-election polling: The prediction of winning for a candidate results in a higher vote for that candidate than would have occurred without the prediction, so long as the predictor is credible. Typically, people want to be on the side of victory, the better side, the effective side, the side with more resources, or if it's junior high school, the cool crowd.

STEM is proving to be very effective, resulting in resource pool expansion. The federal budget included nearly $3 billion for STEM education in 2016. The private sector's annual investment in STEM education is estimated at well over $1 billion annually, with companies like Lockheed Martin and Northrop Grumman each investing over $13 million each year, according to the recent *Washington Post* article "Growing Roots for More STEM." Many states, regional alliances, communities, and individual schools are pumping funds into STEM as well (Censer 2012). That kind of infusion draws attention, expanding the community while stretching its goals and mission. As a problem, the bandwagon effect is a nice one to have—sure beats the alternative. But it can strain brand identity and quality control. The moniker *STEM teacher* brings a certain cache not always anchored to a common definition. Much like the inquiry teaching movement derived from Constructivism that reached a pinnacle with the publication of the National Science Education Standards (NRC 1996), STEM's purveyors can be out ahead of its definition. As revealed in their 2013 article "Inquiry-Based Instruction and Teaching About the Nature of Science: Are

They Happening?," Capps and Crawford found that many teachers self-identify as inquiry adherents when in fact their actions suggest maybe not. Inquiry had a bandwagon problem that diluted the brand, as is a risk for STEM without careful adherence to standards of excellence. Maybe STEM needs to be trademarked.

Trademarking is simple enough—merely file an application with the U.S. Patent and Trademark Office (USTPO). But

STEM, unlike *Bubble Wrap*, *Ping-Pong*, or *Jet Ski*, has acquired widespread public use ahead of anyone claiming a specific, novel reference to her or his concept for the term. In USPTO parlance, that makes it *generic*. Thus, according to the USPTO's guidelines, *STEM* likely cannot be trademarked by anyone. Even if it could be, the possessor of the brand would be responsible for infringement watch. With the ubiquity of STEM, a vast SBI (the *STEM Bureau of Investigation*) would be needed to stay on top of all the encroachment. Others who have fought trademark wars can scare us straight in STEM. The Cinemark movie chain recently sued online children's game Roblox for trademark infringement when it discovered users were building little Cinemark theaters in their online "towns." In another instance, Citigroup claimed that AT&T's use of "thanks" and "AT&T Thanks" infringe on Citigroup's trademarked "thankyou" according to a 2016 Reuters piece, "No Thanks: Citigroup Sues AT&T for Trademark Infringement" (Stempel 2016). And from the sublime to the ridiculous, a dog daycare in Algonquin, Illinois, was sued recently by Starbucks for its choice of name—Starbarks. But of all the trademark suits and squabbles detailed at *SecureYourTrademark.com*, one of the most curious may be Harley Davidson's attempt to trademark the rumble of their product's engine. Second place goes to Kentucky Fried Chicken, which took issue when the trademarked phrase *Family Feast* was used by a British pub to brand their Christmas special. Back in the STEM arena, an award for "inspired teaching" netted Iowa's STEM operators a cease-and-desist letter from an east coast educational consultant who'd managed to trademark those two words in sequence.

Legendary infringement suits over musical tunes, high-end sunglasses, and cell phone technology remind us of how challenging it is to hatch an idea, build a reputable brand, and protect it from abuse. No, STEM is on its own in defining and maintaining a trademark. Jean Moon and Susan Rundell Singer captured this condition in their essay in *Education Week*: "Today, not only do we have numerous definitions of STEM, but we also have branded numerous entities to be STEM councils, STEM schools, STEM networks, and STEM curricular outcomes. Despite the well-intended branding, understanding of the brand itself remains elusive." Although they take a helpful shot at a definition of STEM as "an assemblage of practices and processes that transcend disciplinary lines and from which knowledge and learning of a particular kind emerges" (Moon and Singer 2012). That is a good starting point for branding. Does it pick up STEAM?

Picking Up STEAM?

As STEM picks up steam as an effective educational reform movement, partners come to call. Broadening the stakeholder base is a powerful and often beneficial thing, but sticking to the core mission gets harder. That's when a definition comes in handy. ACT, Inc. served the STEM community an enormous favor recently in defining the STEM fields of study after extensive research and stakeholder commentary. (Chapter 6 features ACT's Steve Triplett commenting on the process.) In its annual Condition of STEM report, the

college admissions testing organization lists all imaginable STEM majors and occupations. They range from core sciences such as Genetics and Meteorology to computational fields including Actuarial Science and Webpage Design, to medical fields like Pharmacy and Athletic Training (representing a break from the National Science Foundation's STEM definition which excludes the health sciences), and engineering pursuits ranging from the traditional Chemical and Mechanical Engineering to the more technical like Automotive Technology and Computer-Aided Drafting (ACT 2016). The report (and its valuable counterpart in the Career-Technical realm—the STEM Career Clusters of the Occupational Information Network—O*Net) provides STEM advocates with boundaries (and sometimes shield) for defining scope. Though not included in the ACT or O*Net roster of career pathways, the Arts community at both the local and national level has arduously driven a STEAM movement whose objectives are to "transform research policy to place Art + Design at the center of STEM, to encourage integration of Art + Design in K–20 education, and to influence employers to hire artists and designers to drive innovation" (see the Rhode Island School of Design's STEM to STEAM website at *www. risd.edu/about/STEM_to_STEAM*). In 2013, a bipartisan Congressional STEAM caucus formed to advocate for policy changes that would encourage educators to integrate arts with STEM curriculum. At the local level, STEAM academies have popped up in Lexington, Kentucky; Cedar Rapids, Iowa; Suffolk, Virginia; Gaston, North Carolina; and every other corner of the continent, along with STEAM camps, courses, and conferences.

Innovation and invention course through the veins of today's STEM *awakening*. Imagination and creativity are the lifeblood of innovation and invention. Artistic and design skills translate imagination into invention, and that makes Art very important to STEM. Important enough for its own letter? The agriculture community may want to thumbwrestle over rights to the A. So may accountants and actuaries. Project Lead The Way's CEO Vince Bertram argued that the STEM versus STEAM argument represents a misunderstanding of STEM as merely a static coupling of four academic subjects rather than "an engaging and exciting way of teaching and learning" (2014). Bertram's thesis is "It's not about adding to the acronym, but instead adding to the relevancy of learning. It's about showing students how technical concepts relate to real-world situations and providing them with hands-on projects and problems that help them apply concepts in a new context. It's about nurturing students' curiosity and helping them develop creativity, problem solving, and critical thinking skills" (2014).

Symbolically it is a tempting appeasement for the STEM community to welcome the A (once the litigants rest on whose A it is). But in the wings await the Humanities community (SHTEAM), the Reading folks (SHTREAM), medical (SHTREAMM), computer (SHTREAMM-C), and bottling it all up, HAMSTER (Humanities, Arts, Mathematics, Science, Technology, Engineering, Reading). On the disciplinary bandwagoning of STEM, Iowa's former director of the state Department of Education, Brad Buck, observed that adding letters eventually gets to what we call *school*.

The core mission of STEM—its trademark, unprotected as it may be—is to problem solve a talent shortage in high-demand occupations including engineering, manufacturing, health, finance, IT, and science. Recalling from Chapter 1, the 2011 McKinsey Global Institute study *An Economy That Works: Job Creation and America's Future* noted that enrollment trends portend that in the next decade the United States will produce twice as many graduates in the social sciences and business as in STEM, and that if nothing is done to redirect the interests of some of those graduates toward STEM we can expect an exacerbated shortage of qualified candidates for technical jobs (Manyika et al. 2011). The STEM mission is strengthened by incorporating art, design, reading, music, and the humanities to be sure, but the brand is defined by its funders—governments and businesses who pay for a specific service to solve an economic problem.

STEM Teachers

Youth strategize blade designs for maximum efficiency in the program KidWind. STEM Teachers exhibit specialized skills and knowledge about conveying concepts such as wind energy.

Someone with the skills and knowledge to prepare and inspire problem solvers headed for high-demand occupations including engineering, manufacturing, health, finance, IT, and science must be a STEM teacher. What are those proprietary skills and knowledge that distinguish the STEM Teacher brand? Something of a STEM Teacher Certificate is offered by more and more colleges, nonprofits, and state agencies that will be taken up in Chapter 7. Later in this chapter STEM teacher Ashley Flatebo boils it down to an ability to devise learning experiences that equip students to solve problems that apply outside of the classroom. As with an inquisitive toddler, that answer begets another: What are those abilities?

Before diving in to that, a point of clarification: The phrase *STEM teachers* can be an aggregator, a grouping, as in the mission statement of the 100Kin10 initiative: "100Kin10 unites the nation's top academic institutions, nonprofits, foundations, companies, and government agencies to train and retain 100,000 excellent STEM teachers to educate the next generation of innovators and problem solvers" (*https://100kin10.org/ 2017*). Or, a group of disciplinary

faculty are often referred to collectively as STEM teachers. "Our STEM faculty are second to none" was a recent proclamation by a Midwestern college dean at a business luncheon I attended. Although it's perfectly legitimate to group mathematics, biology, computer, and other content-area teachers as STEM teachers and faculty, it confuses the brand. STEM purists would find it clearer for the 100Kin10 program to declare ambitions to train and retain 100,000 excellent mathematics, science, technology, and engineering teachers because that would more accurately reflect their partners, including such organizations as the Science and Mathematics Teacher Imperative, the Purdue University College of Education, Project Lead The Way, Stanford's teacher education program, and many others who produce excellent content-area teachers. STEM brand protectors would similarly inquire to the college dean as to whether she literally means that her faculty teach STEM, or sub-areas of STEM such as chemistry, computer programming, and civil engineering. Actual STEM teachers are fairly rare, as will be detailed in subsequent chapters, because there are not many preparatory programs making them, nor is there plentiful professional development supporting them (but both are in rapid growth mode). Those who have worked hard to train as STEM teachers would benefit most by preservation of the STEM teacher brand. And that leads back to the question at hand: What are those abilities that define a STEM teacher?

The practices of teachers of mathematics, science, engineering, and technology are well-defined by their standards as discussed in Chapter 1. Those skill sets and abilities may be summarized as including the capacity to operate a safe, collaborative classroom environment; craft active, hands-on investigative opportunities; assess learning in the moment and adjust instructional tactics on-the-fly; incorporate technologies that enhance the learning experience; arouse curiosities; and really know their subject inside and out, yet make it fun and relevant. That's a tall order that thousands of content-area teachers across the United States fulfill as a matter of professional routine. And it is a stellar starting point for aspirants to the moniker STEM teacher. But an upgrade is required, akin to a custom-ordered car. The Toyota Camry, a reliable brand in the United States that sells in the low $20,000s for the base model, will get you where you need to go in safe comfort, like a solid and dependable physics teacher. However, for a few thousand dollars more, the SLE model can take you faster, cozier, and more stylishly with bigger wheels, a navigation panel, and concert-hall speakers. For the physics teacher to "upgrade" to STEM teacher he or she adds on skills and abilities that are the *brand of the STEM teacher*, including the following (listed *a–g* for later reference:

a. Transdisciplinary purview that embraces a seamless approach to learning where investigations are unbound by categorized domains of knowledge

b. Community problem or issue-based learning agenda driven in partnership with learners

c. Work-world applications of standards-driven curriculum, both locally and globally

d. Product-driven assessment that is authentic and meaningful to the community

e. Career guidance on the local/global job market in the field with the education and training pathways to get there, incorporating career exploration into the curriculum

f. Shared leadership philosophy that distributes decision making to learners and community members

g. Intellectual risk-taking and supported failure as an instructive tool for developing perseverance and confidence

STEM teachers, such as the one pictured, prepare for connected learning from the classroom to community, using exemplary programs such as A World in Motion.

But the upgrade to *STEM teacher* is considerably more involved than clicking menu options on your Toyota SLE. STEM teachers subject themselves to extensive personal and professional development to be career coaches, to know the employment landscape, to expand their content knowledge across domains, to see how industry makes use of their content, to flip learners from consumers to producers, to relinquish some control, and to safety-net young tightrope-walkers of the intellect. At this curious and invigorating moment in education, STEM teacher preparatory models, STEM curriculum development and assessment, STEM teacher professional development offerings, administrative supports for STEM teachers, evaluation of STEM teaching, and even student and community STEM teacher embrace can all be hit and miss. The subject of the next chapter is to explore some exciting developments in STEM learning and assessment. Then Chapter 6 explores the STEM teacher preparation landscape, followed by Chapter 8 on emerging STEM professional development innovations. For now, the STEM teacher brand is distinctive and valuable, and foreshadows what must constitute the brand for STEM classrooms and for STEM schools.

CHAPTER 5

STEM Classrooms

The National Science Teachers Association publishes the newsletter *Science and the STEM Classroom.* Each issue focuses on the STEM behind a popular topic such as climate change, the drug industry, adaptive technologies, and so on. It is loaded with teacher tips for developing students' communications skills, integrating content, involving community stakeholders, and more. The breadth and timeliness of each issue characterize a brave new world of teaching and learning in STEM classrooms. Not only does the style and scope of learning look different in a STEM room enlivening those brand features *a–g*, but the physical space adapts to the needs of STEM teams by connecting to the world beyond via technology, accommodating collaborative work, and decentralizing—not much need for a front of the room.

"Over the last three years we have seen a surge … STEM environments have become a focus of districts across the state" reports Dave Bertlshofer of workspace design company Storey Kenworthy, a leading STEM classroom outfitter profiled later in this chapter (personal communication, June 2016). Of the many models that abound of late, the TILE (Transform, Interact, Learn, Engage) classrooms at the University of Iowa inspire a number of K–12 clones throughout the region. The learning space is designed to promote active learning pedagogies and student collaboration, supported with extensive technology. There is no obvious space that represents the traditional front of the room. Furnishings are designed to promote student collaboration with chairs that are movable and tables and writing surfaces allowing students to work in small groups with ample surfaces for student work (whiteboards, glass boards, or slate boards).

The University of Iowa's TILE classroom in use.

Redesigned Learning Environment for STEM at Hoover High School in Des Moines.

Reprinted with permission from Hoover High School.

The TILE classroom design inspired a Redesigned Learning Environments (RLE) competition to spawn more such STEM classroom spaces throughout the state, incentivized by mini-grants and guidelines from the Iowa Governor's STEM Advisory Council. The mission of the STEM RLE program is the creation of a re-imagined learning space that reflects the latest thinking in both a physical environment for active, investigative, technologically enhanced collaborative learning, and a pedagogical environment supported by an educator fully prepared to facilitate project-based, authentic learning linked to the outside world of STEM professionals and community assets. The STEM RLE was conceived of as a secondary learning innovation but has proven to function well at the elementary- and middle-school levels as well. Three requirements of RLE builders are (1) an innovative physical learning space; (2) integrated STEM curriculum focused on personalized, deeper learning based on real-world challenges and college/career readiness objectives; and (3) genuine community partnerships involving higher education, businesses, community organizations, and nonformal education. All this, on a $50,000 grant from the STEM Council. A total of 13 RLEs operate across Iowa, serving as beacons for the direction of education in the state. A reflection on the potency of a redesigned learning space by STEM teacher and administrator Maureen Griffin of Hoover High School in Des Moines centers on the omnipresent student cell phone distraction as a gauge "You don't ever see a kid grabbing those [their phones] in here. There's no reason to because they're so engaged with what they're doing. They forget about the fact that they need to be on their phone."[1] More about Griffin's STEM school transformation follows later in this chapter.

---

1 Quotes in this section are from the author's personal communications with Griffin in June 2016.

STEM classrooms are branded less by physical design than by the actions that take place within, but powerful anecdotes like Griffin's come frequently from career educators who witness sometimes unforeseen transformations. Another recipient of the STEM Council's RLE grant noted a significant drop in discipline issues after the first year's implementation, when educators were actually expecting more, considering the open environment. Comfortable spaces plus interesting studies are a winning combination.

Vital to the success of STEM classrooms are of course school administrators and colleagues, as well as the prevailing policy environment. STEM teacher Ashley Flatebo, featured later in this chapter as a case study in branding STEM teaching, identified a key attribute of school leaders, "Having administration that allows teachers to take risks and are open to new ideas helps build a school with a growth mindset" (personal communication, June 2016). Colleagues, too, factor prominently in the success of a STEM classroom where the contribution to broader school culture can be revelatory or resisted. Finally, local and state policies that permit or impede integrative STEM courses of study, or that nurture or negate the flexibility needed to truly partner with professionals beyond the school walls factor prominently in the full realization of a STEM classroom.

STEM Schools

Specialty schools in mathematics and science have been around for decades. A leading organization, the National Consortium for Specialized Secondary Schools of Mathematics, Science, and Technology, formed in 1988 as a 15-member community that included the well-known Thomas Jefferson High School for Science and Technology in Virginia, the Illinois Mathematics and Science Academy, North Carolina's School of Science and Mathematics, and the Louisiana School for Mathematics, Science, and the Arts. In 2014, the organization updated its name to the National Consortium of Secondary STEM Schools (NCSSS), and now sports over 100 institutional member schools (along with dozens of affiliates and associates). The NCSSS brand is protected by a rigorous application process

that asks evidence and artifacts of inquirers on track record regarding advanced curricular offerings including AP/IB/dual enrollment opportunities, project/research requirements/ offerings, affiliations with local colleges/universities/research facilities, a list of students' acceptances for colleges and universities, and a letter of recommendation from a distinguished partner attesting to the applicant's success as a community leader in STEM education. That's a tidy testimonial to assuring brand preservation.

But STEM schools are proliferating across the country, with more than a dozen in Ohio, several dozen in Texas, hundreds across states including North Carolina, Oregon, Massachusetts, Virginia, California, New York, Arizona, Utah, Washington, New Jersey, Maryland, Iowa, and more. Some states have strict credentialing mechanisms in place to assure a standard of excellence, and all STEM schools self-impose a rigorous bar of achievement, naturally. But as is the condition among STEM teachers, there is no *SBI* (*STEM Bureau of Investigation*) probing and "busting" schools that fail to measure up. The STEM School brand is vulnerable to wide variation. "Some simply bring a more intensive approach to a traditional curriculum, beefing up the offerings and requirements in science, math, and related subjects," wrote *Education Week*'s Erik Robelen in his prescient 2011 column, "Others emphasize project-based learning and integrating ideas across disciplines. Still others may focus on particular occupational themes, such as biotechnology" (Robelen 2011).

The nonprofit school accreditation agency AdvancEd now offers a review and certification service to schools that validate the quality, rigor, and substance of STEM educational programming, according to its website. STEM programs are gauged against the standard "STEM students have the skills, knowledge, and thinking strategies that prepare them to be innovative, creative, and systematic problem-solvers in STEM fields of study and work." AdvancEd provides a "mark of STEM distinction and excellence for those institutions that are granted the certification" (AdvancEd 2017).

Also certifying schools and districts for their STEM prowess is the National Institute for STEM Education, a division of the Houston-based private company Accelerate Learning. Created at Rice University, the institute's certification "integrates the most recent

research and best practices in STEM, 21st Century Learning and professional development." Schools seeking the $12,500 credential develop and implement a STEM Action Plan that includes teacher professional development and the submission of a portfolio, all intended to recognize "individual school campuses for their commitment to—and growth in—teachers' implementation of 21st Century and STEM strategies" (National Institute for STEM Education 2017).

Would-be certifiers notwithstanding, today's STEM school landscape is something of a wild west, with three broad categories having emerged: the specialty schools that prepare learners for focused career paths, for example Buffalo, New York's, Research Laboratory of Life Science and Bioinformatics; selective schools to which learners compete for seats, for example the highly competitive Thomas Jefferson High School for Science and Technology that admits about 15% of applicants annually; and inclusive STEM schools that behave like public community schools in welcoming all comers. The inclusives are by far the most frequent and therefore the focus of the remainder of this discussion.

Inclusive STEM Schools

Inclusive STEM schools were defined by the 2011 National Research Council report *Successful K–12 STEM Education: Identifying Effective Approaches in Science, Technology, Engineering, and Mathematics*: "These schools seek to provide experiences that are similar to those at selective STEM schools while serving a broader population. Many inclusive STEM schools operate on the dual premises that "math and science competencies can be developed, and that students from traditionally underrepresented subpopulations need access to opportunities to develop these competencies to become full participants in areas of economic growth and prosperity" (NRC 2011).

The panoply of models popping up across the nation are downright inspiring despite their wide variance. Two agencies (of many) that have contributed guidelines for school leaders follow. They align and overlap with each other, and reinforce the seven STEM teacher brand factors *a* through *g* previously discussed. Then two of the most successful U.S. states in creating a consistently high-quality network of inclusive STEM schools will also be profiled in the Case Studies section that concludes this chapter.

STEM School Study

The STEM School Study (S3) at the University of Chicago studied 25 inclusive STEM schools in 2014 and derived eight commonalities across them. In its report "The Eight Essential Elements of Inclusive STEM High Schools," an important upshot was, "instructional practices and culture in these schools are often equally if not more important to their STEM teaching identity" than the content of *S, T, E* and *M*. And, "In many inclusive STEM schools … their self-identification as a 'STEM school' comes more directly from their focus on pedagogy and the school culture" (LaForce et al. 2016). Their eight elements of inclusive STEM schools are as follows:

1. Rigorous Learning—locally generated, real-world curriculum

2. Problem-based Learning—interdisciplinary, student-centered

3. Personalization of Learning—student interests and skills steer instruction

4. Career, Technology, and Life Skills—workplace practices (21st-century skills)

5. School Community Belonging—trust, respect, supportive transitions

6. External Community—students and teachers take learning beyond school walls

7. Staff Foundations (termed a *supporting element*)—collaborative, reflective professional development

8. Essential Factors (also a supporting element)—flexible staff with adequate resources, student body representative of the diversity of the community

Opportunity Structures for Preparation and Inspiration (OSPrI)

A team representing George Washington University, George Mason University, and SRI Education, through funding from the National Science Foundation, created a STEM inventory of 14 critical components that define inclusive STEM high schools from in-depth qualitative studies of 8 exemplar schools. From their 2015 study of critical components, the Opportunity Structures for Preparation and Inspiration project identified these components (excerpted for brevity; learn more at *https://ospri.research.gwu.edu*):

1. "Rigorous courses in all four STEM areas, or engineering and technology, are explicitly, intentionally integrated into STEM subjects and non-STEM subjects in preparation for college."

2. "Opportunities for project-based learning and student production are encouraged, during and beyond the school day. Students are productive and active in STEM learning, as measured by performance-based assessment practices that have an authentic fit with STEM disciplines."

3. "Technology connects students with information systems, models, databases, and STEM research; teachers; mentors; and social networking resources for STEM ideas during and outside the school day. ... The school's structure and use of technology has the potential to change relationships between students, teachers and knowledge."

4. "Learning opportunities are not bounded, but ubiquitous. Learning spills into areas regarded as 'informal STEM education' and includes apprenticeships, mentoring, social networking and doing STEM in locations off of the school site, in the community, museums and STEM centers, and business and industry."

5. "School boundaries extend beyond the school community by creating partnerships with business and industry. The school environment intentionally

reflects the workplace. … Students have the opportunity to interact with industry professionals and present their work in professional venues."

6. "School schedule is flexible and designed to provide opportunities for students to take classes at institutions of higher education or online."

7. "Teachers are qualified and have advanced STEM content knowledge and/or practical experience in STEM careers. The school has opportunities for in-house professional development, collaboration, and interactions with STEM professionals."

8. "The school's stated goals are to prepare students for STEM, with emphasis on recruiting students from underrepresented groups."

9. The school's "administrative structure exhibits an external awareness to the community outside the school that promotes a bias toward innovation and action, while also increasing the collective capacity of the school."

10. "Supports such as bridge programs, tutoring programs, extended school day, extended school year, or looping exist to strengthen student transitions to STEM careers."

11. "Teachers use summative assessments (e.g., mastery based learning) to inform future instruction and to enhance student learning. School leaders and teachers examine standardized and summative assessment data to inform teaching strategies, student supports, professional development opportunities, and resource allocation. … Data are available to students and parents."

12. "The school leadership is proactive and continuously addresses the needs of teachers, students, and the greater community through innovative solutions, open communication, and uplifting leadership."

13. The schools have an "environment where students and staff feel a sense of personal, intellectual, and socio-emotional safety. … Students understand that it is acceptable to make mistakes in the learning process and are encouraged to take intellectual risks. Teachers are also encouraged to collaborate and take risks in innovative instructional practices."

14. "Students choose to attend a STEM-focused high school and understand the challenges that will be involved and develop a sense of purpose coherent with the school mission."

The overlap between these 14 OSPrI "critical components" and the 8 "major Elements" of the STEM Schools Study alongside the 7 STEM teacher brand factors *a–g* referred to earlier in this chapter provide the STEM community with a strongly suggestive, emergent STEM schools brand. It is left to the reader as to whether 7, 8, 14, or another number of criteria says it best.

Case Studies on Branding STEM Teachers and STEM Schools

Reports and studies aside, street-level practitioners wielding the STEM brand have their own interpretations and interests in quality control. STEM educators Ashley Flatebo (in school) and Deb Dunkhase (out of school) offer familiar, trademark-able insights on the characteristics of STEM teachers.[2] STEM school champion Maureen Griffin captures the emergent vision of so many across the country building on the STEM brand. Michael of Texas and Tina of North Carolina lead their states' STEM Schools programs, cracking the kitchen door for a peek into where internationally renowned systems are seasoned, stirred, and baked to perfection. The last word is reserved for designers and architects of the physical environment for STEM learning whose work reveals how the sum of factors contributing to a STEM classroom can exceed the value of each part.

STEM Teachers Define the Brand

For Flatebo, a STEM teacher at Lincoln Intermediate School in Mason City, Iowa, the identity of a STEM teacher is entirely bottled up in the opportunities he or she provides for students. Her school provides sufficient STEM opportunities to have been recognized nationally as a 2015 Outstanding STEM Middle School at the Florida Education Technol-

ogy Conference, for its work integrating STEM throughout the curriculum. The school is known regionally for its after-school robotics club and Family Science Night. Lincoln is a full partner with the Iowa Governor's STEM Advisory Council's Scale-Up initiative (described in Chapter 6), bringing high-quality STEM programs like FIRST Lego League, A World in Motion, and other exemplars to their students. And Lincoln participates in the Council's Redesigned Learning partnership as well. The STEM culture there is alive and well.

Elaborating on her definition, Flatebo incorporates essential elements of STEM schools, "A STEM teacher allows students to solve problems that they can apply outside of the classroom." She firmly establishes student-centeredness, "The students are designing and creating solutions and products that they believe will be a solution to their problem." And Flatebo is an interdisciplinarian willing to distribute leadership: "The activities and projects they delve into are authentic and engaging and can connect to other curricular areas. When you walk into my classroom you will not find me in front of a teacher guided room, you will find students engaged in projects, working at their own pace, able to verbalize the task and

---

2 Quotes in this section are from the author's personal communications with the individuals profiled.

its importance, and you find me providing feedback and questioning students to continue their learning as I move from student to student or group to group."

Flatebo's ability to stay at the forefront is a twist on tradition, but wholly in line with the times, "The most valuable professional development for me is Twitter. I routinely participate in chats and follow other STEM leaders and schools within the state and country. This allows me to share what I am doing and gain ideas from others to enhance the curriculum within my classroom." And as an Instruction Coach, one of Flatebo's responsibilities is to instill a culture for STEM beyond her own classroom. She enlists a corp of ambassadors to help, "Students can drive what happens in a school. Through discussion with their peers and the excitement in their voices, a spark within the school will start." Through them, she wins over the community, "When students go home and can explain their learning to parents/guardians and are excited to share, parents/guardians begin to take interest in what is happening in the classrooms." Up and down the school halls, mutual trust incubates innovation at Lincoln, "Teachers that are willing to open their doors and allow other teachers in and share their ideas will allow others to see the engagement level of students." And none of what a school like Lincoln accomplishes is possible without supportive leadership, "Having administration that allows teachers to take risks and are open to new ideas helps to build a school with a growth mindset."

Deb Dunkhase is an out-of-school STEM educator, directing the Iowa Children's Museum (ICM) in Coralville, Iowa. The ICM is an award-winning mecca of exploratory discovery for youngsters and their families from across the Midwest, and a shining light of innovation on the power of play. Dunkhase and her staff of "Play-ologists" adhere to a STEM educator definition remarkably in line with Flatebo's as well as the guideposts portrayed earlier, seeing their role as "helping students explore a problem or challenge using science and math concepts to build their understandings of the issue followed by the use of engineering and technology in search of a solution to the problem or challenge." And that's

often toddlers flexing their mental muscles. The informal learning environment is especially well-suited to be active and child-centered, according to Dunkhase. "STEM education becomes powerful when content is presented in the context of a real-world situation." A welcome and recurrent theme. Dunkhase continues, "When the four disciplines of STEM are fully integrated into the student's everyday life experiences,

it matters more to both the teachers and the students." Content and interactive learning experiences at the ICM are all built around content that is interdisciplinary and recognizable to the learner.

For play-ologists at the ICM, "It's all about going where the kids are and providing a basic map of discovery for them to follow with the knowledge that the students are going to take numerous twists and turns along the way," exemplifying the essential element of flexibility. By respecting and valuing those detours, "that's where the kids are going to find the self-confidence and STEM identity they're going to need in life to be successful." To bring a high-quality standard of STEM excellence to the ICM, Dunkhase and her team committed to a professional development model that equips them to design and assess programs rather than students, a key differentiator from the formal school milieu. The *Dimensions of Success STEM Program Quality Assessment Tool*, developed by the PEAR (Program in Education, Afterschool and Resiliency) Institute at Harvard University guides the ICM in comprehensively examining its STEM practices, from how it uses space to how students interact with exhibits and each other, as well as what content and reflections arise as outcomes (Program in Education, Afterschool and Resiliency 2016). Dunkhase closes with a signature of the STEM brand—continuous improvement through collaboration, "Our staff are committed to high-quality STEM education but no matter how well we serve the kids with whom we work, we welcome the opportunity to share and learn from other like-minded STEM informal educators across the state."

STEM School of Distinction: Hoover High School, Des Moines, Iowa

How does a school go about establishing itself STEM in the hearts and minds of those who teach at, attend, and surround it? In states like Texas and North Carolina where agencies and systems are in place to exert control over the label, there is a process discussed shortly. But for many, it is a matter of internal decisions and actions that hopefully match up with known best practices. Hope has become reality at Hoover High School in Des Moines, Iowa, where several factors over the course of four years contributed to their STEM school transformation.

"Our journey began with a district level decision to change school-based management teams from the traditional principal/vice principal structure to include school improvement leaders [SIL] who are subject-specific administrative team members," recalled Maureen Griffin, former science teacher and now the SIL responsible for developing the Hoover STEM Academy. Coinciding with that mission came a grant for a Redesigned Learning Environment from the Iowa Governor's STEM Advisory Council. "This was a transformative process for our district as everyone from the superintendent, district technology team, furniture design team, school principal and STEM team leader came together to work together to transform the RLE spaces in the school." A year later, a private-sector grant secured similar learning space at the middle school for the introduction of coding and robotics at the eighth-grade level. In true entrepreneurial spirit, Hoover leaders secured one

more potent grant from the STEM Council to support a STEM BEST (Business Engaging Students and Teachers; see Chapter 4) project to partner with area universities and businesses, providing students with internship experiences during their senior year of high school. Meanwhile, additional private funding expanded computer and robotics programs at the high school, accounting for what might be characterized as an accretion approach to STEM transformation by leveraging and maximizing resources.

Today, Hoover's STEM Academy exists within the walls of a comprehensive high school. "STEM students can still be in band, orchestra, and drama or compete in a sport as a Hoover Husky," Griffin clarified, "but they also benefit from the programming and opportunities of the STEM program," which include a weekly Speaker's Bureau, Genius Hour, Scientific Research course and field trips, internships, and perhaps most important, a personal connection: Each student is matched with a teacher who serves as advisor for all four years of his or her high school career and gets a dedicated administrator who supports their educational and socio-emotional needs as they navigate the high school experience. For their part, teachers within the STEM Academy undergo professional development focused on interdisciplinary approaches to teaching and learning, STEM team identity development, and effective classroom practice. "It is important to note," shared Griffin, echoing the upshot finding by SRI about STEM schools identities, "that our STEM teachers are receiving added learning and training because they are science, technology/engineering, and math teachers; but in reality, all Hoover students benefit from the engaging strategies our teachers are learning and practicing daily."

The calling card for any STEM school, by Griffin's measure, is "a clear and intentional focus on developing the 'innovators mindset' in students." It is a network of mutually supportive teachers and professionals connected to the broader community along with learning environments thoughtfully accommodating of such mindsets that make all the difference.

Standardized Systems of STEM School Excellence: North Carolina and Texas

Several U.S. states have instituted systems of STEM school recognition. Two have led the movement with thoughtful and innovative mechanisms for framing, vetting, supporting, and thereby brand-protecting a network of inclusive STEM schools, North Carolina and Texas. A classical instance of parallel evolution, architects of the two systems developed sets of rigorous criteria, established supportive structures, and grapple with similar challenges born of success. Tina Marcus is project manager for STEM Education and Leadership at the North Carolina Department of Public Instruction, and she boils down the work of NC STEM as providing a GPS waypoint to guide schools aspiring to the STEM brand. In Texas, Dr. Michael Odell is vice president for research and technology transfer and professor of STEM education at the University of Texas at Tyler. He co-founded several of the state's STEM schools designated T-STEM Academies along with a supporting T-STEM Center (one of seven) that provide curriculum, professional development, and community advocacy.

"There are over 154 schools designated as Texas STEM Academies as of May 2016," according to Odell, adding "Keep in mind there can be STEM focused schools that do not seek designation."[3] Similarly in North Carolina, Marcus points out that there are "many outstanding schools and programs in our state executing exceptional STEM strategies producing pronounced results. However, for schools to attain and be honored as 'STEM Schools of Distinction' they go through an application process …" to be described shortly, with result that "to date, we have identified 23 STEM Schools of Distinction that encompass public and charter, elementary, middle, and high schools; K–8 and early college high schools."

Both Texas and North Carolina mapped out the key attributes that unite all of their STEM schools under a common, rigorous banner. According to Odell, "The cornerstone of T-STEM Academy learning is student engagement and exposure to innovation and design in STEM-focused instruction and learning that models real-world contexts." Standards of excellence guide each of the T-STEM Academies in the form of a Design Blueprint, Rubric, and Glossary "as a guidepost to build and sustain T-STEM schools that address the seven benchmarks: (1) mission-driven leadership; (2) school culture and design; (3) student outreach, recruitment, and retention; (4) teacher selection, development, and retention; (5) curriculum, instruction, and assessment; (6) strategic alliances; and (7) academy advancement and sustainability."

North Carolina's STEM Schools of Distinction must meet 11 essential attributes measured by a STEM Attribute Implementation Rubric. Those 11 attributes are contained in three overarching principles:

1. The curriculum should be content driven and integrated across subject areas (inquiry based, problem based, and project based) that are real-world and connect to current and relevant issues occurring in our society and industries.

2. The school should have ongoing community and industry engagement. This engagment should include applied learning experiences for teachers and work-based experiences for students with external industry partners either during the school day, outside, or in informal environments.

3. The school should have connections to postsecondary institutions, aligning student career pathways with strategies that include diverse course selections and career explorations.

Once a school attains STEM status in both states, renewal and sustenance become important. In Texas, redesignation is an annual process by which schools are gauged against a rubric calibrating their successes in advancing on the seven benchmarks. North Carolina's Marcus expects "schools having attained the STEM School of Distinction designation to reapply three years after having received the designation." Given the depth and thoroughness of the NC STEM application and review process, Marcus's team believes that

---

3 Quotes in this section are from the author's personal communications with the individuals quoted.

"allocating a period of three years will allow schools time to reflect and chart new strategies as they continue to grow their STEM education program."

Maintaining favorable status once designated can sometimes present its own challenges. Marcus points out that "staff and administrative turnover can be a challenge," and so, too, can school master schedules be an inhibitor in terms of "carving out a common time for collaboration and planning." Similarly in Texas, Odell observes that administrator turnover can present a sustainability challenge that sometimes may cause, while in other instances may contribute, to the struggle seen in some schools to maintain fidelity to the Blueprint principles. Back in North Carolina, Marcus has witnessed a recurrence of the sustenance philosophy of STEM school as recorded by the STEM School Study of exemplars, "The feedback we have been receiving from schools using the implementation rubric is that if

you take the word 'STEM' out of the implementation rubrics, you still have the best school improvement tool."

How do these leading thinkers frame the future of STEM schools? Both Odell and Marcus represent organizations that have produced national models so that one outcome of their statewide work may be to expand the exemplary mechanism for standardized quality in STEM school designation nationwide. Texas has an advantage in "very liberal dual credit rules," that factor in to the success of the T-STEM Academies, to the point, by Odell's estimation where "we think that the College Freshman Year may go away as it will be completed in high school." Likewise a trail is being blazed in North Carolina, where "the schools implementing great strategies of STEM education will be valuable and constructive magnets attracting, creating, and growing robust communities-at-large and a win–win for the community and economic development." Marcus looks at STEM as a connector rather than an island, "No one entity can do this alone and STEM education is not a stand-alone."

STEM Schools—The Physical Environment: DLR Group and Storey Kenworthy

Learning spaces, as any teacher knows, can make a significant difference in attitudes of students. And attitude can make or break the effectiveness of a lesson. Fortunately there are professionals who have made it their business to design, build, and furnish habitable STEM learning environments. The recipients of Re-designed Learning Environment grants and STEM BEST grants of the Iowa Governor's STEM Advisory Council often turn to two providers of state-of-the-art educational spaces—the DLR Group, an architecture,

engineering, planning, and interiors firm in Des Moines, Iowa, and Storey Kenworthy, a statewide leader in innovative workplace solutions.

"We believe design can elevate the human experience" said DLR's Nick Hansen, Business Development Leader. Storey Kenworthy's Account Manager Dave Bertlshofer concurs, "We believe the learning environment truly impacts the level in which students can learn." The best partners for schools in redesign are the ones who exhibit such passion, but also understand the pedagogical landscape. Bertlshofer goes into partnerships with an opening question, "How can we design a space that best improves the communication and collaboration among students and teacher?" And his counterpart Hansen similarly understands where STEM education is trying to go, "Ideal spaces are those that are responsive and nimble, adapting to the needs of current learners and educators, as well as those who will learn and educate in the future."

Design inspiration may come from an obvious but sometimes undertapped source. "Some of our most innovative designs come from the students," said Bertlshofer, "We witness school leaders actively listening to the feedback their students provide, and in most cases they implement those ideas in some degree into the space." Over at DLR Group, a similar page in the playbook describes how "Some of our most important sources for design inspiration are the students and educators themselves—observing, interviewing, and brainstorming with them ignites our creativity." Bertlshofer is quick to credit designers Ken Hagen and Andrew Van Leeuwen for that part. Both firms recognize the ultimate buy-in of teachers and administrators to be paramount, actively seeking input and guidance every step of the way.

In these budgetary times of austerity (as if there's been an alternate condition in education!), Bertlshofer of Storey Kenworthy recommends not to skimp on choices in design, "When given the opportunity to choose where and how they want to learn, students are more focused and productive." Hansen at DLR Group establishes the must-haves when funds are tight by working with the client on key guiding principles that define the non-negotiables. Safety and security, well-placed exits, climate control, and natural daylighting—aspects that directly affect the well-being of students and staff—are all nonnegotiable.

Despite lean school finances, both Bertlshofer and Hansen observe a rapid uptick in the demand for learning-space redesign. "Over the last three years we have seen a surge, an increase of energy and excitement in education. STEM environments have become a focus of districts across the state often as a starting point in creating a higher standard of learning," Bertlshofer shared. Hansen and the DLR Group, too, acknowledged hundreds of recent renovation and retrofit projects for aging Iowa school facilities. "Over the last five years there has been an increasing demand for learning environments that challenge preconceived ideas of what a learning environment should be," he said.

As competitors in an exciting growth industry, innovative education design firms must always be looking ahead. Where these companies see trends headed are toward ever more

flexibility in concept and design. "The learning spaces of the future are student-centric and technology rich," prognosticated Hansen, "These flexible environments encourage collaboration, critical thinking and group work, and meet the unique needs of students, educators and the community as a whole." Bertlshofer looked *Back to the Future*, quoting Dr. Emmet Brown from the 1985 movie of that title, "Where we're going, we don't need roads." Brown elaborated that the education environment is forever changing, and "it is necessary to continue to push our own limitations and creativity to work to provide the best learning environment for the students and instructors."

The STEM Brand: A Familiar Melody

A recent lawsuit claimed that the legendary rock band Led Zeppelin plagiarized from a lesser known band the opening chords of their biggest hit, *Stairway to Heaven*. The judge who eventually dismissed the charges decided that the melodic style was a musical tradition that dates back at least to the 1600s. No one could "own" it. STEM is education's *Stairway to Heaven* and there are symphonies performing similar tunes across the nation. Although no one entity may claim sovereignty of the brand *STEM Classroom*, *STEM Teacher*, or *STEM School*, a consensus on essential features—whether they be called attributes, elements, characteristics, components, or benchmarks—has clearly congealed. Play on, STEM bandwagon.

References

ACT. 2016. Condition of STEM report. ACT. *www.act.org/content/act/en/research/condition-of-stem-2016.html*.

AdvancEd. 2017. STEM certification. *www.advanc-ed.org/services/stem-certification*.

Bertram, V. *Huffington Post*. 2014. STEM or STEAM? We're Missing the Point. May 26.

Capps, D. K., and B. A. Crawford. 2013. Inquiry-based instruction and teaching about nature of science: Are they happening? *Journal of Science Teacher Education* 24 (3): 497–526.

Censer, M. *Washington Post*. 2012. Growing Roots for More STEM. April 22.

Henshel, R. L., and W. Johnston. 1987. The emergence of bandwagon effects: A theory. *The Sociological Quarterly* 28 (4).

LaForce, M., E. Noble, H. King, J. Century, C. Blackwell, S. Holt, A. Ibrahim, and S. Loo. 2016. The eight essential elements of inclusive STEM high schools. *International Journal of STEM Education*. doi 10.1186/s40594-016-0054-z.

Manyika, J., L. Lund, B. Auguste, L. Mendonca, T. Welsh, and S. Ramaswamy. 2011. An economy that works: Job creation and America's future. McKinsey Global Institute. *www.mckinsey.com/global-themes/employment-and-growth/an-economy-that-works-for-us-job-creation*.

Moon, J., and S. R. Singer. 2012. Bringing STEM into focus. *Education Week* 31 (19): 24, 32.

National Institute for STEM Education. 2017. National Certificate for STEM Excellence: Campus certification—Support and requirements. *http://acceleratelearning.com/nise/resources/ nise_campus_district_certificate_program.pdf*.

National Research Council (NRC). 1996. *National science education standards*. Washington, DC: National Academies Press.

National Research Council (NRC). 2011. Successful K–12 STEM education: Identifying effective approaches in science, technology, engineering, and mathematics. Washington, DC: National Academies Press. *www.stemreports.com/wp-content/uploads/2011/06/NRC_STEM_2.pdf*.

Program in Education, Afterschool and Resiliency (PEAR). 2016. Dimensions of success: A PEAR observation tool. PEAR. *www.pearweb.org/tools/dos.html*.

Rhode Island School of Design. 2017.

Robelen, E. 2011. Latest wave of STEM schools taps new talent. *Education Week* 31 (3): 1, 18–19.

Stempel, J. *Reuters*. 2016. No Thanks: Citigroup Sues AT&T for Trademark Infringement. June 10.

CHAPTER 6

STEM Learning and Assessment

Identifying and developing top STEM curriculum, implementing that curriculum with high fidelity, and assessing what really matters as the outcome of strong STEM curriculum and pedagogy—the three-legged stool on which excellent STEM teachers lean. After all, interwoven with classroom and teacher designations are curricula and assessment strategies. Exemplary materials that support STEM teaching and learning are plentiful, as are emergent processes of homegrown school–business learning platforms. Investigative, interdisciplinary, and applied learning definitive of STEM means a flipping of assessment—both what is assessed, and how. Vanguard thinkers in this space are creatively dovetailing local and state-mandated assessments with authentic measures.

STEM Gets "the Bends"

A classic STEM lesson was dealt to humanity in the middle 19th century, when a disease surfaced which was not caused by malnutrition or a genetic mutation, nor a trans-species virus or other contagion, but by technology. Decompression sickness, or "the bends," showed up in the 1840s when engineers first began digging very deep holes to access coal beds or to anchor bridge foundations to bedrock. In his 1988 book on the subject, *The Bends,* John Phillips documented how steam engines introduced compressed air technology which in turn opened up the depths (and heights for that matter) to human exploration. Rich coal veins beneath the Loire River in France could be accessed by sinking a long steel box called a *caisson* (envision a vertical tunnel) to the river bottom, pumping out all the water, then sending workers down to harvest coal in dry comfort. But to keep the water out, the chambers had to be sealed tight and pumped with two or three times the air pressure of the surface. Workers started suffering a variety of ailments ranging from hearing loss to bloody cough, headaches, and sometimes far worse. But symptoms often passed and were infrequent enough that progress prevailed. By the 1860s, caisson technology had been imported to St. Louis for building the foundations to the Eads Bridge across the Mississippi River. It is there that the 60-foot deep compressed air work environments yielded bent-over workers at shift's end, their joints aching and limbs partially paralyzed. Locals assigned them the bends, though the cause was still a mystery. At the maximal depth of 90 feet, workers began dying rather frequently. Not until the 1880s were nitrogen bubbles in the blood pinpointed as the cause.

When it comes to STEM learning and assessment, the bends provide both an illustrative case study and symbol of our progress all in one. As an instructional centerpiece, decompression sickness couples the applied worlds of engineering and technology with the pure sciences of chemistry and physiology. It's a prime candidate for three-dimensional learning as prescribed by the *Next Generation Science Standards* (*NGSS*): the *science and engineering practices* (developing models, investigations, analysis, computation, and explanation) are woven throughout; the *crosscutting concepts* (mechanisms, systems, proportions, stability, and change) are inherent; and *disciplinary core ideas* (matter, molecules, Earth systems, and engineering design) abound. The topic has broad applications to air travel, ocean exploration, mountain climbing, and even cooking. It is relevant to any swimmer and bridge-crossing hodophile. Not only does it transcend *S-T-E-M* but the construction engineering context invokes economics, sociology, and history (Who were those desperate workers? What economic forces trumped safety? Where was OSHA?). As career guidance goes, the bends lends seamlessly to job explorations in respiratory therapy, oil rig construction, aerospace engineering, meteorology, and more. That is STEM learning, in a nutshell. The remainder of this chapter expounds on key characteristics of STEM learning experiences akin to Chapter 5's elements of STEM teaching. Spoiler alert: they overlap.

The other half of this chapter explores STEM assessment, invoking the bends as a symbol of our progress on that front. STEM is as much an educational innovation as was the caisson an engineering innovation. Both enable deeper exploration, but both are incompletely perfect. The caisson created access to depth but had not accounted for the effect on people's health. STEM opens up deep and broad three-dimensional learning but assessment systems may not always align with goals and outcomes of STEM classrooms. Tests and grading can cause a mental case of the bends for STEM learners and their teachers. But like the physiologists at the turn of the 20th century who figured out time tables and gas mixtures for minimizing the bends, innovative STEM educators have solved the assessment challenge of answering two masters: the diagnostic and prescriptive assessment needs of learners, and the external constituent's need for grades and test scores.

STEM Learning

When it comes to STEM, *how* you teach is more important than *what* you teach. That maxim weaves throughout the *NGSS* and the *Common Core State Standards for Mathematics*. For example, "the focus is on the core ideas—not necessarily the facts that are associated with them. The facts and details are important evidence, but not the sole focus of instruction" (NGSS Lead States 2013). And "across the … mathematics standards, skills critical to each content area are emphasized. In particular, problem-solving, collaboration, communication, and critical-thinking skills are interwoven into the standards," respectively (NGAC and CCSSO 2017). Entire books have been written of similar title, such as Nancy Sulla's 2015 *It's Not What You Teach, but How.* It's a well-worn conclusion that surfaced in the find-

ings of the STEM Schools Study of Chapter 5; it is what Marcus's clients discovered in North Carolina; it occurred to Mason City's STEM teacher Ashley Flatebo and Des Moines Hoover's STEM leader Maureen Griffin as they carved out their STEM missions: STEM learning is branded less by content coverage than by the actions that take place.

The *what* of STEM learning is well defined by the *NGSS*, *Common Core State Standards for Mathematics*, the standards for technology from ITEEA and ISTE, and the engineering education objectives embedded in *NGSS* and other existent standards. (In a related aside, the National Academy of Engineering studied the feasibility of engineering education standards but opted instead to embed them into existent disciplinary standards out of a realization that there would be significant barriers to introducing another set of standards to the already burdened K–12 curriculum [see NRC 2010].)

The *how* of STEM learning is its *sine qua non,* its non-negotiable brand definition. This book opened with and has since continuously reinforced a consensus definition that if it is STEM taking place in a classroom, club, girl scout troop, day care, museum, or family kitchen, it's active, collaborative, interdisciplinary (or *trans*disciplinary), content-rich in applications, contextualized to community connections, personally meaningful to learners, risk friendly, and career compassed.

Knowing *what* makes up a STEM learning experience and *how* it is taught and learned is like knowing what a car is and how it operates. There remains seeing it in action, doing the driving. Some of the STEM drivers introduced already include Chad Janzen with Rocket Manufacturing in Chapter 2, where students apply their knowledge to crafting metalwork solutions for local manufacturers. And Ehren Whigham in Chapter 4 whose students brew up root beer in partnership with the Confluence company. Chapter 5 drivers were Ashley Flatebo whose students practice authentic and engaged learning; Maureen Griffin and her genius hour and personal connections; Deb Dunkhase helping youth navigate twists and turns, and Michael Odell of Texas and Tina Marcus of North Carolina with their benchmarks and attributes, respectively, steering STEM schools across regions. Illustrations of STEM learning in action abound. Here come a few more.

Sample of Top STEM Programs

There's no match for homegrown STEM programs that connect learners to the issues of their communities, whether the community be defined as local or global. For the *edupreneur* (an educator who designs and launches innovative learning solutions), establishing the partnerships and building the curricula that can meet the STEM threshold for *how* can be a happy task. Chapters 2 and 3 mapped the process. But not every teacher self-identifies as an edupreneur, according to Aaron Tait and Dave Faulkner in their pithy reader *Edupreneur: Unleashing Teacher Led Innovation in Schools* (2016). It may be that an educator has yet to develop the skills of bridge-building, external networking, or cross-stakeholder curriculum development. In some instances, educators find themselves operating in a school leadership environment that inhibits rather than incubates edupreneurialism. Fortunately in such instances, there is a nearly-as-good-as-homegrown option—abundant and outstanding STEM programs that own the *how* of STEM learning.

The Governor's STEM Advisory Council in Iowa has become intimately acquainted with some of the best-known STEM programs through an initiative that vets, markets, distributes, and evaluates courses and curricula for scaling across the state, hence the title *Scale-Up.* Chapter 1 referenced these "pre-packaged" STEM programs, and Chapter 4 credited Scale-Up programs as partnership starters. Scores of nationally recognized programs as well as a few locally developed ones have been scaled to thousands of educators

and hundreds of thousands of Iowa youth to date. They apply for scaling in Iowa by providing evidence of alignment with the best practices of STEM, namely that they are active, collaborative, interdisciplinary, content-rich in applications, contextualized to community connections, personally meaningful to learners, risk friendly, and career compassed. Some of those programs with descriptions pulled from their websites are as follows:

- **A World in Motion** (AWIM; *http://awim.sae.org*). Developed by the Society of Automotive Engineers, AWIM is a teacher-administered, industry volunteer-assisted program that brings science, technology, engineering and math (STEM) education to life in the classroom for K–8 students. Benchmarked to the national standards, the AWIM program incorporates integrated STEM learning experiences through hands-on activities that reinforce classroom STEM learning.

- **Engineering is Elementary** (EiE; *www.eie.org*). Developed by the Museum of Science, Boston, EiE is the nation's leading engineering curriculum for grades 1–5. This fun, flexible, inquiry-based curriculum gives you everything you need to integrate engineering with the science subjects you already teach. Engineering calls for children to apply what they know about science and math. Their learning is enhanced as a result. And because engineering activities are based on real-world technologies and problems, they help children see how disciplines like math and science are relevant to their lives.

- **HyperStream** (*http://hyperstream.org*). Developed by the Technology Association of Iowa to help develop Iowa's future IT workforce, HyperStream connects grade 5–12 students with industry professionals who expose the students to potential IT careers within Iowa, as well as mentor the students as they complete IT-related curriculum and projects. The program provides students with access to technology-based education and real-world experiences. HyperStream focuses on five learning tracks rooted in computer science: Multimedia, Game Design, Cyber Defense, Application Development, and Robotics.

- **Project Lead The Way** (PLTW; *www.pltw.org*). is a nonprofit organization providing transformative learning experiences for K–12 students and teachers across the United States PLTW empowers students to develop in-demand, transportable knowledge and skills through K–12 program pathways that include PLTW Launch (K–5), PLTW Gateway (6–8), and high school programs: PLTW Biomedical Science, PLTW Engineering, and PLTW Computer Science. PLTW's teacher training and ongoing support mechanisms allow teachers to engage their students in real-world learning in schools, preparing them to seek rewarding careers, solve important challenges, and contribute to global progress.

- **Spatial-Temporal (ST) Math** (*www.mindresearch.org/stmath*). Developed by MIND Research Institute, ST Math is game-based instructional software for K–12, designed to boost math comprehension and proficiency through visual learning, to ensure all students are mathematically equipped to solve the world's most challenging problems while developing perseverance and problem-solving skills, and becoming life-long learners prepared for success.

- **KidWind** (*www.rechargelabs.org*). This is a program to bring effective renewable-energy STEM training to educators through its REcharge Labs and provide students with hands-on applications of their knowledge with the KidWind Renewable Energy Festival and the Online Renewable Energy Challenge. These events allow students to share their projects in a supportive and engaging environment. REcharge Labs cover the integration of wind and solar topics into grades 2–12 and the use of all kits and materials.

- **Curriculum in Agricultural Science Education** (CASE; *www.case4learning. org*). CASE develops curriculum utilizing science inquiry for lesson foundation and concepts are taught using activity-, project-, and problem-based instructional strategies. In addition to the curriculum aspect of CASE, the project ensures high-quality teaching by providing extensive professional development for teachers that leads to certification. Curricular materials provide a high level of educational experiences to students to enhance the rigor and relevance of agriculture, food, and natural resources (AFNR) subject matter. Besides elevating the rigor of AFNR knowledge and skills, CASE provides purposeful enhancement of science, mathematics, and English language understanding.

- **FIRST Tech Challenge** (FTC; *www.firstinspires.org/robotics/ftc*). Developed by For Inspiration and Recognition of Science and Technology, FTC is a mid-level robotics program designed to inspire and increase the interest of young people (ages 12–18) into STEM fields. FTC offers students the opportunity to design, build, and program robots; build experience and confidence with complex STEM-based concepts; document the engineering design process; develop problem-solving and team-building skills; enhance their public speaking skills; and compete and cooperate in alliances during tournaments. Additionally, FTC enables students, including those traditionally underrepresented in the STEM fields, to solve real-world challenges and offers a life-changing experience to help students realize a STEM career is feasible.

- **Pint-Size Science** (*www.sciowa.org/learn/pint-size-science*). Developed by the Science Center of Iowa, Pint Size Science introduces preK and kindergarten children to STEM topics through discovery learning. Using a hands-on approach that engages and inspires young minds to explore scientific phenomena, the program works to not only build science understanding but also respond to the ever-changing interests and abilities of children.

- **Making STEM Connections** (*www.sciowa.org/makingstemconnections*). Also developed by the Science Center of Iowa, this curriculum framework is focused on

the idea that making and tinkering are ways to engage a student's mind and build conceptual understanding around academic content. Making STEM Connections is structurally supported by cross-curricular experiences and opportunities, including literacy and math, to reinforce the maker's foundation of active learning and problem solving. The purpose of making as a learning technique is summed up by Dale Dougherty, chairman of Maker Education Initiative, "It is the difference between a child who is directed to perform a task and one who is self-directed to figure out what to do" (Dougherty, n.d.).

- **SEPUP** (Science Education for Public Understanding Program; *http://sepuplhs.org*). Units support the development of core science content and the practices of science and engineering through a variety of hands-on learning activities. Issues related to global sustainability in the life sciences serve as themes for the core ideas in the *NGSS* and provide continuity and thematic connections in each unit. The program features high-quality, hands-on materials and support for the development of scientific and language literacy, and a nationally recognized, rubric driven assessment system.

Every SEPUP unit uses personal and societal issues, such as global threats to human health, sustainable agriculture/ecosystem management, and maintaining biodiversity and more, to provide thematic continuity and motivation for student investigations.

The organization Change the Equation operates a national process to identify exemplars that is akin to Iowa's Scale-Up called STEMworks (see Chapter 2). It is a searchable database of in-school and out-of-school STEM programs that have been independently evaluated against a rigorous rubric, providing a reliable roster of dozens of proven winners that educators and funders can bank on to drive STEM learning (2017). To be scaled in Iowa, a program must apply for STEMworks first and be rated at the "promising" or "accomplished" levels.

Others in the business of reviewing and rating STEM programs include the New York Academy of Sciences (NYAS) and the National Commission on Teaching and America's Future (NCTAF). The NYAS operates a free STEM education certification program that essentially coaches curriculum developers, teachers, policy makers, and other STEM consumers on high-quality products. "Countless organizations are developing STEM content, resources, and instructional programs for the education sector," notes its website (*http://globalstemalliance.org/program-areas/certification*), "Yet little objective information exists to help decision makers identify high-quality materials." Programmers may apply to the NYAS's Global STEM Alliance for certification, and to date, three programs have been recognized by "a panel of experts with advanced degrees and deep knowledge related to the subject matter, grade level, and intended audience of the materials under review."

The NCTAF offers a slight twist on certification. According to its website, it provides to curriculum coordinators, instructional coaches, and teachers "a set of design and evaluation tools, protocols, rubrics, and case studies to guide and monitor the complex, interdisciplinary curriculum and assessment work that the new standards demand" (see *https://learningstudios.nctaf.org/#!/rubrics*). NCTAF's STEM Learning Studios integrate three strategies—collaborative teacher teams, external STEM partnerships, and project-based learning. Free design, monitoring, and evaluation tools provide ways to gather data and provide productive feedback to teacher teams.

And speaking of evaluation, all of this STEM learning calls for a different lens on assessment.

STEM Assessment

STEM educators beware of a mental case of the bends over an assessment challenge of dual purposes: diagnostic and prescriptive assessment needs of learners, and the external constituent's need for grades and test scores. The good news is that pioneers on this front have championed high-quality STEM assessments while proving the efficacy of STEM teaching at raising grades and test scores.

An Assessment Reckoning

Harried educators rarely get an opportunity to step back and take a circumspect, high-altitude view of the profession. Assessment calls for just that. Consider all of the other things that people assess daily, sometimes unconsciously: the health of a tree in the backyard; the fashion sense of the barista at the coffee shop, the dog's IQ, the value of an $8 car wash, the medium-ness of a grilled steak, the neighbor's rendition of James Taylor's "You've Got a Friend," or the passage of time documented by the mirror. It is all a matter of judgement. Every assessed instance requires the judge to tap a mental scale ranging from good to bad—a rubric in the head. The quick cerebral calculus generates a snap opinion bordering on truth—that tree is drought-stricken; that guy can sing! Then comes an action that alters the future course of the thing observed—a compliment reinforces the barista's affinity for magenta hair; the $10 car wash might better remove that tar; more sunscreen. There's an alignment and a linearity to assessment in real life—something of value is observed, judged on a relied-on (and hopefully reliable) scale from good to bad and all points in between, and

acted on to move it ever more in the direction of good. Cycle repeats. STEM assessment calls for such alignment: learners intake, process, and do things, then they and their mentors judge the observable against a scale (the learner's own emergent as well as the mentor's more reliable), followed by actions that move the observable aspect of the in-taking, processing and doing of STEM ever more toward good.

Aligning Assessment With STEM

The *what* of STEM is vastly easier to assess than the *how*. It is quicker, more objective, and readily reportable to determine to what degree learners can balance chemical equations, compute a trig function, or recite the string variable types in Java. However, the world agrees that *how* is more important than *what* in STEM learning, so there is no choice but to take up the more time-consuming subjective assessment approaches that measure learners' abilities to use and apply STEM. "This new vision of science learning presents considerable challenges—but also a unique and valuable opportunity for assessment …. Existing science assessments have not been designed to capture three-dimensional science learning, and developing assessments that can do so requires new approaches" (NRC 2014).

The *NGSS* are loaded with *performance expectations* that describe what learners ought to know and be able to do as a result of three-dimensional learning, at each grade level. That gives educators a great advantage in assessing skills, processes, and content gained. But structures still need to be created by educators to assure that assessments abide by learning theory while validly measuring what's valued and addressing practical restraints such as time and resources. The NRC report on developing assessments includes numerous examples of individual and class-wide assessments that capture a comprehensive picture of learner gains in practices, core ideas, and crosscutting concepts. For example, learners' understandings of biodiversity are assessed by a teacher prompt to determine the number of species in the school yard, to represent the finding graphically, to derive from the data a response to the initial prompt, and to aggregate findings into a comprehensive biodiversity-by-region explanation for the school. What this and all of the examples of the report have in common is that students "need to design an investigation, collect data, interpret the results, and construct explanations that relate their evidence to claims and reasoning" (NRC 2014).

A cavalcade of commercial and nonprofit curriculum developers and vendors strive to provide *NGSS*-reflective assessment tools to STEM educators. The nonprofit Measured Progress struck a recent partnership with Activate Learning to produce middle school–level assessments of their *Investigating and Questioning Our World Through Science and Technology* curriculum so that "instead of amassing science information, students apply their understanding to new situations that require them to think critically, reason intelligently, and make sense of science as they engage in practices such as developing models, explaining phenomena, and using evidence to argue in support of their claims," according to Activate Learning's website (*www.activatelearning.com/about-us*). The elementary-level engineering platform Materials World Modules proclaims, "Evaluation and assessment is an integral part of The Materials World Modules Program" aligned to the *NGSS* through "content and methodology [that] seamlessly integrate the three dimensions—Practices, Cross-Cutting Elements, and Content—outlined in the *NGSS*, with a strong emphasis on Engineering Design" (see *www.materialsworldmodules.org*). The curriculum products companies and organizations LAB-AIDS, Carolina, Discovery Education, the Smithsonian Science Education Center, STEMscopes, Lawrence Hall of Science, Pearson, and nearly countless more have each carefully aligned learning and assessment products with the performance expectations of the *NGSS*.

Personal, homegrown examples of STEM assessments are in the "Profiles in STEM Learning and Assessment" section to come. But first, a visit on the challenge of systemic STEM assessment in an era of accountability buffeted by resource constraint.

STEM Assessment for External Constituents

Like tectonic plates grinding in opposite directions, the desire for classroom, school, and education system accountability bumps up against limited resources, emitting tremors across the landscape. It is not so much a problem of misalignment between what is taught by day in a STEM classroom and what gets tested on state or national assessments, but an issue of fit. Despite mighty efforts on the part of organizations such as the National Assessment of Educational Progress and the International Baccalaureate program to embed hands-on challenges into science assessments, the practice has not caught widespread use. Inevitable resource challenges, such as costly kits and intensive processing expense, inhibit a best practice. The result is that what can be asked and easily scored *en masse* does not always fit with what goes on in STEM classrooms. The NRC observed that in an ideal world of course, assessment tasks at the classroom level and at the system level would look a lot alike, but standardized tests pose additional challenges: "They need to be designed so that they can be given to large numbers of students, to be sufficiently standardized to support the intended monitoring purpose, to cover an appropriate breadth of the *NGSS*, and to be feasible and cost effective for states." (NRC 2013). Until such a test comes along, educators may take solace in findings that suggest STEM learners may do just fine on standardized assessments anyhow.

A case in point is the Iowa Assessment—a standardized statewide test administered annually to nearly every student in grades 3–11, about 370,000 youth. Students are assessed on both mathematics and science as well as vocabulary, reading, social studies, and other skills/knowledge. For the last few years the test has also included an optional subset of attitudinal questions regarding STEM interest in cooperation with evaluators of the Governor's STEM Advisory Council. The math and science scores as well as attitudes of Iowa youth who do and do not take part in Scale-Up programs of the STEM Council are able to be compared annually, and are compiled in a yearly assessment report produced by the Center for Social and Behavioral Research (CSBR) at the University of Northern Iowa.

About 2,000 educators take part in Scale-Up programs each year, delivering an exemplary STEM program to approximately 100,000 preK–12 youth. A total of 990 educators responded to the most recent CSBR evaluative survey in which 77% reported observing increased interest in STEM among their students. According to the CSBR, educators also "thought that students' critical thinking, problem solving, confidence, and collaboration skills were developed or showed improvement throughout the program" and that "students' critical thinking, problem solving, confidence, and collaboration skills were developed or showed improvement throughout the program." Respondents reported their students

showing "more perseverance in their problem solving and would try harder to find solutions before giving up or asking for help" and observed students "thinking more like scientists and engineers." Finally, educators observed their students "applying math, science, and technology to real-world problems and wanting to bring the ideas they learned into their communities." These excerpts of the 2015–2016 annual report, along with all evaluative archives are available at *http://iowastem.gov/iowa-stem-evaluation* (see Heiden et al. 2016).

Of all Iowa Assessments takers across the state, 29,396 were able to be identified as having participated in STEM Scale-Up (they were the participant pool of the 990 educators who uploaded the requested information). Two-thirds of them were elementary level, about one-fourth were middle school, and one-tenth secondary. That is important because the council's elementary programs are almost entirely delivered to "involuntary" groups of youth, that is, to classroom teachers, as opposed to the secondary programs that are elective courses or self-selective clubs. In contrasting the Iowa Assessment scores of STEM program participants against their peers, the evaluators found the following (Figure 6.1, p. 112):

- Higher interest across the STEM spectrum of subjects

- Higher interest in STEM careers

- Higher interest in working in Iowa upon completion of studies (a new item of curiosity for the council, accommodated by an added question on the Iowa Assessments)

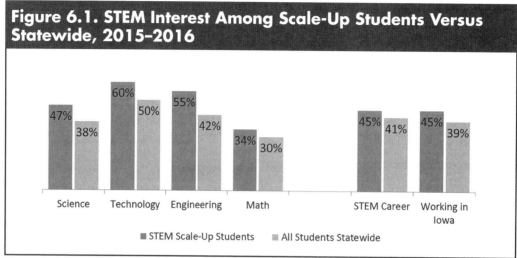

Figure 6.1. STEM Interest Among Scale-Up Students Versus Statewide, 2015–2016

Source: Iowa Governor's STEM Advisory Council; reprinted with permission.

More important for the purpose of this chapter, evaluators also compared content scores between Scale-Up participants and their statewide peers. They found that STEM students scored an average of 7 percentage points higher in National Percentile Rank in mathematics, an average of 6 percentage points higher in science, and an average of 4 percentage points higher in reading (an unanticipated but welcome finding).

The effect was true across age groups and particularly pronounced at the elementary level (Figure 6.2). And the gain for minority students was higher yet: +10 in mathematics and +8 in science compared with minority students statewide.

Maybe STEM educators can have their cake and eat it, too. Not only can they conduct active, collaborative, interdisciplinary, applied, content-rich, community-connected, personally meaningful, risk-friendly, and career-compassed classes for learners, they can also be preparing learners to thrive on standardized exams, based on the experience of Iowa.

Profiles in STEM Learning and Assessment

Organized by dimension, the following practitioners of STEM learning and assessment explicate the alignment process in real time working with teachers and students. High school technology educator Ben Kuker incorporates summertime industry experiences into learning and assessment tasks. Then agriculture educator Melanie Bloom dives deeper into the Iowa Scale-Up partner CASE (Curriculum in Agricultural Science Education) as a STEM learning and assessment exemplar. Finally, the American College Testing Program's (ACT's) STEM assessment provides a Mount Everest purview of U.S. graduates' interest and readiness. Steve Triplett goes behind the scenes on development of the ACT STEM suite of solutions.

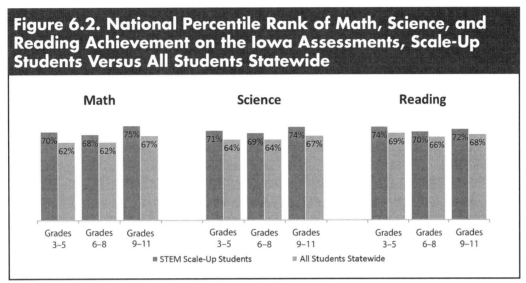

Figure 6.2. National Percentile Rank of Math, Science, and Reading Achievement on the Iowa Assessments, Scale-Up Students Versus All Students Statewide

Source: Iowa Governor's STEM Advisory Council; reprinted with permission.

Homegrown STEM Curriculum

American Profol, Inc. is a manufacturer of films used in wraps, packaging, adhesives, graphics, and other almost unimaginably diverse applications. The company takes summer time teacher externs in partnership with the Iowa Governor's STEM Advisory Council. One of those recent externs was Ben Kuker, a technology educator at Mt. Vernon High School. He found a great match between his Introduction to Engineering course and a problem dealt with at American Profol. "Production operators, maintenance technicians, and management continually try to increase the efficiency of production," he had come to learn over summer, though challenges arise when switching equipment over from one batch to another. "For instance, if one customer orders yellow plastic film but the next order calls for blue film, how can the company reduce the amount of wasted material and employee hours switching production orders?"[1] A class project was born.

Kuker engaged his students in the situation by producing a video that explained the cast film production process at American Profol—a career guidance opportunity itself. It was carefully framed in production and equipment terms that freshman and sophomores could appreciate. All of the information students would need was embedded, including links and resources that would assist students as they started gathering information and contemplating ideas. The conclusion of Kuker's video challenges students to develop a solution that would increase the efficiency and reduce the down time when changing production orders. They love an open-ended, real life challenge.

Like all STEM educators in standard K–12 environments, Kuker strives to be creative in innovating while meeting the school's and state's curricular requirements. "It is so easy to fall into a pattern of teaching the same information, in the same manner, and anticipating

---

1 Quotes in this section are from the author's personal communications with Kuker in June 2016.

Technology educator Ben Kuker (pictured) spent six weeks one summer applying his skills in industry at American Profol. He then translated the experience to authentic lessons for students.

the same results," the veteran teacher has found, but the result would nag him in that "then all we are asking our students to do is perform at the same level that the past class has reached." Kuker expects more of students and of himself. "I make sure to cover what is required and what many would consider important. Though by varying the delivery of the lessons I can include the same concepts but present them in varying mediums." As an example, the American Profol challenge replaced an established activity that focused on developing 21st-century skills in learners. He just found a better, newer way. And with a powerful twist: "At the completion of my design challenge students were required to present their solutions to the president and plant manager of the business. The hope was for everyone within the board room to accurately understand what the student solution was and how it solved the production efficiency problem." Now that is applied community relevance and an authentic assessment.

Kuker's assessment entailed providing students with the required components that they were to account for in their slide presentation: design brief on the problem, research and generating solutions brainstorm, solution development rendition, prototype construction, document results and recommendations, and presentation. When it came to grading though, "This was the toughest part of the activity," Kuker reflected, since there was no history to draw from. He envisions refining a rubric scored by the degree of completion/depth for each area. STEM assessment is ever-evolving.

A CASE of STEM Learning and Assessment

CASE was profiled earlier as one of the "off the shelf" STEM exemplary programs available to educators across the nation. It is project- and problem-based, focusing on the rigor and relevance of agriculture, food, and natural resources subject matter, integrating science, mathematics, and other disciplines throughout. Melanie Bloom is an agricultural science educator with more than a dozen years of experience, a CASE Master Teacher, and a member of the CASE writing team.

The CASE connection to STEM is seamless and research-based. "CASE activities, projects, and problems fit into the Daggett Rigor/Relevance Framework" said Bloom,

referring to Willard Daggett's 2005 International Center for Leadership in Education paper "Achieving Academic Excellence Through Rigor and Relevance."[2] Accordingly, the CASE lesson development philosophy is that a well-designed lesson or course "provide the knowledge base, technical skill, and eventually the cognitive reasoning required for solving complicated real life problems," as quoted from the CASE Lesson Development Philosophy of 2012 (see *http://bit.ly/2oeOdhz*). The excerpt goes on, "By providing this balance of experiences, students are prepared for challenges of postsecondary education and future careers." To equip teachers for executing CASE with high fidelity, Bloom cites rigorous professional development: "Our lead teachers spend 60–80 hours modeling instructional practices for participants to emulate and add to their teaching toolboxes."

Like Ben Kuker, Bloom and CASE teachers confidently innovate by integrating local and state curriculum-aligned materials and activities with novel approaches to learning and assessment. Again she anchors CASE classroom practice to the CASE Lesson Development Philosophy, "Core academic standards are used in the development of lesson concepts when natural connections with science, mathematics, or English language elements are present in the context being addressed." The life science focus of CASE readily crosswalks to the *NGSS* disciplinary core ideas of energy, ecosystems, and traits; science and engineering practices such as models, investigations, communication; and naturally crosscutting concepts including scale, proportion, and systems. Moreover, CASE learning experiences range from "activities using structured inquiry, projects relying on guided inquiry, and open inquiry represented with problem-based learning experiences," according to Bloom.

When it comes to assessment, Bloom cites the CASE Assessment Philosophy and Strategies protocol of 2013, which states, "CASE curriculum uses a blend of formative and summative assessment strategies [and] … employs several strategies to ensure that student misconceptions are not present, or proficiency of skills and knowledge is learned before moving the student to the next set of information" (CASE 2013). A number of assessment tools are provided to educators for gauging student growth in the moment, through a lesson, and over time. Flexibility is essential for innovative STEM educators assessing sometimes on-the-fly or first time through as in Kuker's case. Bloom and CASE recognize this and build in the flexibility for teachers to "modify, adjust, create, and mold the curriculum to fit local needs. This includes assessment." In fact, she elaborated, "We have teachers across the country utilizing CASE assessments in traditional grading, standards-based grading, and point system environments alike."

The Mount Everest of STEM Assessments: ACT

The highest vantage across the land for viewing STEM learner trends may best be provided by ACT. Since 2013, the nonprofit assessment/research organization has delivered a monumental service especially to the STEM community: *The Condition of STEM*, an annual report on the emerging STEM education pipeline at both the state and national

---

2 Unless otherwise noted, quotes in this section are from the author's personal communications with Bloom in June 2016. The Daggett paper Bloom discusses is available at *www.leadered.com/pdf/Achieving_Academic_ Excellence_2014.pdf*.

levels. Combining well-established and regarded subject area assessments in science and mathematics with ACT's Interest Inventory, the annual Condition of STEM report is a valuable snapshot on college readiness in these important fields. ACT's support of the nation's STEM imperative has expanded to include a robust and authoritative definition of STEM majors and occupations, a STEM-specific rating on the ACT test, the STEM College Readiness Benchmark, and an ACT Aspire STEM score for grade 3–10 learners who take part in ACT's Aspire assessment (not to mention the ACT WorkKeys and the ACT National Career Readiness Certificate available to learners and employers to assist in determining STEM-related job readiness). For the purpose of this chapter, however, the focus is on ACT's college-readiness STEM assessment.

The ACT STEM score comes from students' mathematics and science scores and according to Steve Triplett, formerly ACT's STEM Initiatives program director, it represents a student's overall performance in these subject areas. Mathematical reasoning skills for solving problems in the areas of algebra, geometry, and trigonometry, combined with science reasoning skills of interpretation, analysis, evaluation, and problem-solving across life, physical, and Earth sciences make up the STEM score. A value-add to the ACT STEM score is the subsequently developed STEM Benchmark for indicating college readiness in STEM. "Students enrolled as STEM majors take different first-year math and science courses than those taken by most college students," points out Triplett. "The STEM Benchmark is based on students' chances of success in Calculus, Chemistry, Biology, Physics, and Engineering courses. An ACT STEM score of 26 is associated with a 50% chance of earning a B and about a 75% chance of earning a C in these entry-level STEM-related courses."[3] The STEM Benchmark is associated with meeting the challenges to persist all the way to a STEM-related college degree.

A student's STEM score combined with revelations of the ACT Interest Inventory provide a powerful window into an individual's STEM trajectory. "The Interest Inventory consists of six scales with 12 items per scale," Triplett explains. "The two scales relevant to the assessment of STEM are *science and technology* and *technical*. The Science and Technology Scale references investigating and attempting to understand phenomena in the natural sciences through reading and research and in applying such knowledge. The Technical Scale focuses on working with tools, instruments, vehicles, and mechanical or electrical equipment."

Leaning back from the metric details, Triplett takes a big-picture approach to ACT's involvement in STEM, "As the lack of U.S. students interested in STEM fields was beginning to become problem, members of the research team led by Jill Crouse and Matt Harmston began to focus on the need for a report that could provide hard data about the state of STEM readiness." The development of the STEM Report was going to depend on identifying students who had an interest in STEM. Already ACT collected students' responses to the planned major/occupation questions asked on the registration form. "Then statistical analyses revealed that, in addition to major/occupation choice, two ACT Interest

---

3 Quotes in this section are from the author's personal communications with Triplett in June 2016.

Inventory scale scores (science and technology) showed up as important factors in determining STEM interest," said Triplett, lending to a set of variables that permitted the definition of two types of interest—expressed and measured. And that in turn led to three STEM interest student groups—expressed and measured, expressed only, and measured only. "Once this system of identifying STEM interest was in place," says Triplett, "the report could be designed to contain information that was important for stakeholders and useful for improving the state of STEM readiness in the United States." That was 2013, when the concept was presented to a focus group of STEM education leaders who invited ACT to consider disaggregating the data by ethnicity, educational aspiration, highest parent educational level, and gender for even greater power. Thus, ACT's first STEM condition report came to be in February 2014 and annually since each November in slightly adjusted forms.

The generation of definitive STEM majors and occupations was a critical contribution by ACT. Of 294 major and occupation titles that registrants choose from, 93 were identified as STEM. An expert panel of ACT researchers knowledgeable of labor market trends and postsecondary academic programs were informed by three sources of information in making the determination: (1) STEM-designated occupations from the U.S. Bureau of Labor Statistics (BLS), (2) STEM-designated degree programs from the U.S. Immigration and Customs Enforcement (ICE), and (3) interest inventory score profiles for students planning to enter the major/occupation. To be classified as STEM in the array of ACT major and occupation titles, the major had to either be categorized as STEM by both the BLS and ICE or else by the BLS plus the occupation had a high interest frequency among students whose profile of measured

interests peaked on the science and technology scale. "These two guidelines accounted for 89 of the 93 ACT titles that were assigned to STEM," recalled Triplett. "The remaining four titles were assigned to STEM based on the judged intensiveness of their math and science coursework (major) or work tasks (occupation)."

STEM stakeholders use *The Condition of STEM* in a variety of ways. First and foremost, students and their parents can rely on the tool to help them align expectations with future course demands. Teacher, counselors, and administrators can better understand the majors and occupations of interest to their graduates and use this data to better inform current and future students on academic priorities, preparation, and goals. At the postsecondary level, faculty, advisers, and administrators can use information in the STEM condition report to better understand the challenges STEM students face so as to offer supports that improve perseverance in STEM majors. At the policy level, programmers and funders have a potent gauge of interest and readiness to help evaluate efforts. And employers have a window into the pipeline trends affecting talent availability and recruitment. From here, Triplett notes that beginning in 2016, ACT will add the STEM Score and STEM Benchmark as reporting categories when students get their ACT results.

STEM Cures "the Bends"

STEM has come a long way since the post-Sputnik Science-Technology-Society (S-T-S) movement spurred a new brand of interdisciplinary, applied, connected learning, and assessing. What inhibited the S-T-S movement according to Robert Yager in Chapter 1 (lack of instructional resources and an over-value on basic recall) has been resoundingly overcome in the 21st century through exemplary curricular products, enlightened mass assessments keying on processes and analyses, and foundational guideposts including the *NGSS*. The advancement of learning and assessment in STEM, like grasping the physiology of pressurized depths a century past, is still evolving toward full alignment. The STEM experience focused on *how* more than *what* translates to performance-based three-dimensional demonstrations of achievement not easily captured by mass quantity measures. Yet, STEM learners can rise to the occasion, excelling on wide scale standardized assessments as well. Kuker and Bloom live it daily. And ACT captures a 20,000 foot snapshot of the STEM learning and assessment landscape each year for us.

References

Change the Equation. 2017. STEMworks database. *http://changetheequation.org/stemworks.*

Curriculum for Agricultural Science in Education (CASE). 2012. CASE lesson development philosophy of 2012. *www.case4learning.org/images/documents/CASE%20Lesson%20 Development%20Philosophy.pdf.*

Curriculum for Agricultural Science in Education (CASE). 2013. CASE assessment philosophy and strategies. *www.case4learning.org/index.php/assessment-learning-reflections/philosophy-and-strategies.*

Dougherty, D. n.d. The maker mindset. Massachusetts Institute of Technology. *https://llk.media. mit.edu/courses/readings/maker-mindset.pdf.*

Heiden, E. O., M. Kemis, M. Whittaker, K. H. Park, and M. E. Losch. 2016. *Iowa STEM monitoring project: 2015–2016 annual report.* Cedar Falls, IA: University of Northern Iowa, Center for Social and Behavioral Research.

National Governors Association Center for Best Practices and Council of Chief State School Officers (NGAC and CCSSO). 2017. Frequently asked questions: Do the *Common Core State Standards* incorporate both content skills? *www.corestandards.org/faq/do-the-common-core-state-standards-incorporate-both-content-and-skills.*

NGSS Lead States. 2013. *Next Generation Science Standards: For states, by states.* Washington, DC: National Academies Press. *www.nextgenscience.org/next-generation-science-standards.*

National Research Council (NRC). 2010. *Standards for K–12 engineering education?* Washington, DC: National Academies Press.

National Research Council (NRC). 2013. Developing assessments for the *Next Generation Science Standards.* Report Brief. *http://sites.nationalacademies.org/cs/groups/dbassesite/documents/webpage/dbasse_086205.pdf?_ga=1.225339098.1762565096.1486052812.*

National Research Council (NRC). 2014. Developing assessments for the *Next Generation Science Standards.* Washington, DC: National Academies Press.

Phillips, J. 1998. *The bends.* New Haven, CT: Yale University Press.

Sulla, N. 2015. *It's not what you teach, but how.* New York: Taylor & Francis.

Tait, A., and D. Faulkner. 2016. *Edupreneur: Unleashing teacher led innovation in schools.* New York: John Wiley & Sons.

CHAPTER 7

The Making of STEM Teachers

Many preK–12 teachers of mathematics, science, engineering, and technology teach STEM. But chances are they came about it the hard way by value-adding their knowledge and skills beyond disciplinary collegiate preparation. New teacher credentialing programs in STEM are emerging. How they go about making a STEM teacher who can craft and assess a STEM lesson within a STEM classroom of a STEM school networked to the broader community is the question at hand. This chapter delves into preparatory programs and pathways that integrate the best practices illuminated up to this point into the doctoring of a professional STEM educator. Of all the dimensions of STEM education reform, new teacher production may be the most delicate front for its institutional constraints and market forces. That makes it a most important front as well, where innovators are taking on the challenge.

CHAPTER 7

Doctoring a New STEM Teacher

The parallel evolution of physician training and STEM teacher training is a lesson in convergence. Like wings beget flight whether butterfly, bat, or bird, interdisciplinary training begets doctors and STEM educators. Harvard Medical School recently announced major curricular revisions to its doctor-making program. And from the description of the changes in the college newspaper *The Crimson*'s story "Top Medical Schools React to Harvard's Curriculum Change" they might as well be talking about teacher preparation: [The change] "integrates multiple disciplines into single courses and introduces earlier clinical immersion and flipped classrooms" (Fu and Joung 2015). A medical college dean commented in the piece that there is a "shift in medical education from asking students to memorize material to emphasizing its application." The change is said to excite medical students while generating some anxiousness among medical faculty. The dean said they're moving more toward "a way to develop critical thinking more than just doing an information download" (Fu and Joung 2015). How? First, introductory courses such as Foundations of Medicine that integrate topics from pathology and microbiology are replacing discrete courses in those disciplines. Second, traditional lectures, long the mainstay of medical

education, are being "flipped" to use class time for discussion and analysis of video content assigned as homework. And third, medical students are applying what they learn right away in early clinical experiences. By all appearances, the medical training community has read *How People Learn* (see Chapter 1). Switch in "STEM" for every mention of "medical" in this passage and it holds to be equally true.

Doctoring a new STEM teacher calls for a similarly dramatic shift in preparatory practices. Stand-alone courses in physics, mathematics, programming, and the like, along with "Methods in [content area] Teaching" evolve to more thematic titles such as "Global Issues in STEM" and "Methods in STEM Teaching." Instructional practices that model STEM learning environments are the new normal. And early, frequent experiential learning applied to real problems and challenges carry the day. New STEM teachers, in parallel with new physicians, can think and teach more critically, bring cross-disciplinary richness to investigations, and unite learning with doing for and with the broader community.

STEM Credential Programs: *Caveat Emptor*

Medical school reform is rapidly spreading. Duke University, Stanford, Johns Hopkins, Mayo, Baylor, and many more have recently retooled practices akin to Harvard's overhaul. The sweeping changeover is driven no doubt by the market awaiting graduates because hospitals and clinics need a new brand of MD for evolving patient care environments. But that's where the analogy breaks down because STEM teaching license and endorsement demand on the open market is tepid, but warming.

Shoppers for and consumers of STEM educator credentials soon realize that preparatory programs at the undergraduate level or for first-time license-seekers at the postsecondary level, earning the stamp of "STEM-qualified" on a *bona fide* certificate is tough. Burgeoning STEM specialist or endorsement add-ons to existing teaching licenses at universities across the nation portend an inevitability of greater access, but their variations favor a *caveat emptor* approach on the part of credential-seekers, much as for STEM classrooms and STEM schools as discussed in Chapter 5, and STEM curriculum of Chapter 6. The hunt for best practices–based STEM credentialing programs winds through a carnival midway of variants calling out STEM. In his trailblazing 2013 book *The Case for STEM Education*, Rodger Bybee identified nine different perspectives on the use of the STEM acronym, ranging from a singular discipline such as mathematics being identified by its purveyor as STEM, to an exclusive view of STEM restricted to transdisciplinary courses (such as a seminar on sustainability). Many disciplinary programs incorporate some if not most of the features supportive of STEM education (especially science education programs anchored to the *Next Generation Science Standards* [*NGSS*]). Yet there remains a grand opportunity born of confusion about STEM, for the community to come together around defining what a high-quality credentialing program should look like.

STEM Credential "Critical Leverage"

Countless science teachers teach STEM. The same holds true for mathematics, technology, and engineering educators. Most have prepared themselves through professional development (Chapter 8), while others had the good fortune of a science (or other discipline-specific) preparatory program that incorporated the hallmarks of STEM teaching and learning (from Chapter 5): integrating career guidance, focusing on community-based issues and problems for context, blurring lines between the disciplines, applying learning to the work world, generating authentic products as assessment, and safely supporting failure as an instructive tool. The trainers of STEM teachers, too, are a special lot.

The professionals who prepare teachers of STEM adopt a reflective lens trained on the environments and expectations that await their charge. Not unlike architects of medical training who in the overhaul process must step back and ask some very basic questions about how their preparatory model matches the profession that awaits, faculty orchestrating the design of STEM licensure and endorsement pathways must ask the following:

- How can coursework better model active, applied, and collaborative pedagogy?

- What constraints inhibit the incorporation of product- and project-based assessment?

- How can the barriers to interdisciplinary instruction be dismantled?

- Where in the preparatory sequence might career coaching be integrated and modeled?

- When and how could links of coursework to the world beyond best take place?

A mighty opportunity for teacher preparatory institutions is to prepare their own faculty to respond to these needs, in order to deliver to the inevitable market teachers with STEM teaching capacities. And to credential them accordingly.

In his aforementioned book, Bybee brought half a century's accrued wisdom to bear on this matter. He believes that undergraduate teacher education reform insofar as STEM teacher preparation is concerned, can and should use credentialing as "critical leverage." "No discussion of improving STEM education escapes acknowledging the need to change teacher education" (Bybee 2013). His case for reform was preceded by *Preparing Teachers: Building Evidence for Sound Policy* (NRC 2010), a report of the Committee on the Study

of Teacher Preparation Programs in the United States. The committee recalled the origins of formal teacher preparation in the United States: "Beginning in 1879, colleges and universities also started to appoint special professors of pedagogy" (NRC 2010). This trend carried the day and prevails as standard fare. A significant challenge or opportunity for the STEM education community, but especially for teacher preparatory institution leaders, is to imagine what a special professor of STEM pedagogy should know and be able to do, and then make hires or offer training so as to implement the imagined state.

Current mathematics and science teacher-preparers need special professional development in order to prepare STEM educators. The committee pointed out that "the primary document that guides departments of mathematics regarding the teaching of mathematics to future teachers" (NRC 2010) is a 2001 report, *The Mathematics Education of Teachers* by the Conference Board of Mathematical Sciences. In "STEM years," that predates the Big Bang. Over in science education conversely, the challenge is less a matter of dated guidelines but instead scattershot practices. The *Preparing Teachers* report concluded, "We could find no systematic information on the content or practices of preparation programs or requirements for science teachers across states" (NRC 2010). And to Bybee's point, the report continued, "We found very little information about how states are using their authority to regulate teachers' qualifications or the characteristics of teacher preparatory programs, but the hints we could find provided little indication that they are taking full advantage of this authority" (NRC 2010). State-level engagement in STEM credentialing is timely.

One more argument for an authoritative, public agency–certified STEM credential comes from the organization 100Kin10. First mentioned in Chapter 5 as a case-in-point for the aggregate use of the term STEM, the folks at 100Kin10 contributed a valuable bullseye for the collective STEM community to shoot at—a set of Grand Challenges that impede efforts to get excellent STEM teachers into every classroom. Many involve teacher preparation. Paraphrasing from their post at *https://100kin10.org/approach*, their argument is as follows:

- New teachers are underprepared to teach STEM subjects, which is due in part to the lack of effective modeling by professors, a lack of exemplary classroom models, and a lack of expert coaches in STEM.

- STEM content and pedagogy are not integrated through the preparatory experience.

- Traditional school models discourage innovation.

- Schools are disconnected from communities and industries they ultimately serve.

Certificates, endorsements, licensure, and other forms of credentialing at the institutional, state, and someday maybe the national level which can guide and support yet hold accountable organizations that prepare future STEM teachers are Bybee's critical leverage.

STEM Credentialing Programs: STEM Add-on and STEM Original

A survey of existing STEM credential programs necessitates a division: the far more common post-bachelor's certificates, endorsements, and master's degrees that further train licensed educators (the *STEM add-on*) and the rarer first-time license to teach that anoints new professionals in STEM education (*STEM originals*).

STEM Add-ons: Inservice Credentialing

Something of a STEM Teacher Certificate is offered by more and more colleges, nonprofits, trade groups, and even industry partners. Four illustrative examples survey the scene.

1. **Massachusetts STEM Certificate Program** for K–12 teachers (*www.stem certificate.org*) is funded by and operated by PTC, a software technology company in Needham, Massachusetts. Educators from other states applying to enroll in the PTC program, although credential recognition is a matter of local

negotiation. PTC is approved as a professional development provider in the state of
Massachusetts and their certificate verifies the completion of about 67 hours of professional development. Through webinars, an online collaborative community, and summer face-to-face, educators are treated to "training [that] brings a unique industry perspective and is grounded in project-based and interdisciplinary learning." Through project-based curriculum design, teachers learn to implement a process called Explore-Create-Share.

2. **Endeavor STEM Teaching Certificate** (*www.us-satellite.net/endeavor*), funded by NASA and administered by U.S. Satellite Laboratory, is a 100% online K–12 program. "Change in teacher practice is the key to improving student performance in STEM area disciplines" says its website. A three- or five-course Certificate in STEM Education from Teachers College, Columbia University is bestowed on completers. Options are a beginner track that is "often for elementary generalists" and an advanced track "for experienced content experts or other K–12 STEM educators." There are master's degree and six- to nine-month certificate options.

3. **Virginia Tech's Integrative STEM Education graduate certificate** (*http://bit.ly/2peg8T5*) is an online program for educators with the goal "to develop 21st-century P–20 STEM educators, leaders, scholars, and researchers prepared as catalysts of change for teaching, disseminating, and investigating integrative teaching/learning approaches to STEM education," according to the website. The integrative focus is promoted as unique. It is a 12-credit-hour sequence of four courses spanning STEM Foundations, Pedagogy, Trends and Issues, and Readings in Technology Education.

4. The aforementioned **National Institute for STEM Education (NISE)** of the private company Accelerate Learning that offers school certifications also certifies teachers in STEM (*http://bit.ly/2p2Z9n0*). The NISE STEM certificate can be an entrée to a master's and doctorate degree with a partner institution, the for-profit American College of Education. It involves producing a portfolio that demonstrates proficiency across 15 "Teacher Actions essential to STEM learning," including connecting learning outside the classroom, addressing student misconceptions, and developing engineering solutions.

STEM Originals: Preservice Preparation

It is a far heavier lift to mold a STEM educator from raw talent than to value-add a licensed teacher, especially at the undergraduate level. The term "institutional constraint," which was coined by Nancy Brickhouse and George Bodner in their 1992 *Journal of*

Research in Science Teaching report "The Beginning Science Teacher: Classroom Narratives of Convictions and Constraints," to label familiar inhibitors of classroom innovation such as 40-minute class periods and lock-step curriculum applies to teacher preparation equally well. A number of institutional constraints may hamper the establishment of a STEM teacher program at colleges and universities, mostly brought about by competition with established programs in mathematics, science, and technology education. New interdisciplinary content and methods courses must be created. Faculty capable of teaching STEM must be freed up from current duties. Field experiences in STEM classrooms must be identified. Skeptics need to be assuaged that sufficient content across the S-T-E and M equip a teacher for any and all. Meanwhile, beyond the institution, the state's accrediting agency must establish and prescribe a STEM license. And then, there must be a market for the product. Thus, *STEM originals* are rare. Two examples show the way.

The University of Wisconsin at Platteville

The University of Wisconsin at Platteville has a new STEMteaching major for grades 1–8. The university's School of Education STEM Team, coordinated by Assistant Professor Erin Edgington, attributes the innovation to the university's chancellor, Dennis Shields, who charged the School of Education with developing the major in line with "meeting the mission to serve the Tri-State area."[1] There was plenty of motivation on the STEM-focused campus, fueled by data collection: "Students' performance in preK–12 education in the areas of mathematics and science have been bleak—on both international measures such as TIMSS, as well as national measures such as NAEP and the ACT," Edgington acknowledged on behalf of her team. The UW-Platteville School of Education STEM team is committed to improving STEM learning design, in part by creating a degree "that would offer opportunities for students to obtain an MC-EA license (middle-childhood/early adolescence) from the [WI] Department of Public Instruction" while working closely with faculty in STEM "to discuss content considerations in the design of our program," said Edgington.

Inventing a new major from scratch provided Erin and the team with unique insights they gladly passes forward. For one thing, their advice to others is to determine what makes a STEM program "STEM." "The biggest hurdle during the planning phase was how to integrate engineering into preparation for grade 1–8 educators," Edgington recalled, adding, "Initial challenges were in speaking a common language as well as forging new collaborations that had not previously existed with allied faculty and within the School of Education."

The School of Education team happened upon a challenge that will be familar at this point to readers of this book. When researching models, Erin related on behalf of the team "we were surprised to find that most of the programs offered were master's level certifications or certificates." So they built their undergraduate model from scratch through

---

1 Quotes in this section are from the author's personal communications with Edgington in July 2016.

researching STEM broadly and deeply "while attending to recommendations from organizations such as the U.S. Department of Education, American Association for the Advancement of Science (and Project 2061), National Science Teachers Association, STEM Education Coalition, and National Science Foundation, to name a few. We also consider the *NGSS* important for guiding instructional decisions around coursework content."

The vision that guides the UW-Platteville STEM team is grounded in a belief that "teachers trained to design, implement, and measure integrated and project-based learning, while being engaged in learning that models this design, interpret teaching and learning differently" said Edginton. An important difference that uniquely qualifies them with "an intentionality in designing student engagement in a STEM learning model that is inherently different than traditional instructional models."

In planning the STEM teaching major at UW-Platteville, surveys were sent to nearby teachers and administrators to gain perspective. The vast majority of both groups supported the new licensure option. The work has spawned an increase in the number of inservice educators taking advantage of the courses as well, with some pursuing STEM licensing. Edgington and the team foresee a continuous feedback loop of data collection, analysis, redesign as needed, and implementation to a level of excellence befitting a model for other state institutions and beyond. Once the undergraduate program is humming, her team has its sights set on a STEM master's program for inservice teachers. In the interim, pragmatism prevails, "In order for all of this to come true, we will need continued financial support from the state, continued support and collaboration on campus, engagement at the state and national levels in STEM education, and the perseverance to bring our conceptualization of excellent STEM education preparation to fruition."

Iowa STEM Teaching Endorsements

Iowa has had STEM teaching endorsements since 2014, when a committee of the Governor's STEM Advisory Council worked with the Board of Educational Examiners to carefully craft three robust pathways, undergraduate or graduate. The endorsements are for K–8 STEM, 5–8 STEM, and K–12 STEM specialist. The committee was led by then-professor of science education at Central College Dr. Kris Kilibarda, who has since taken up the role of state science consultant for the Iowa Department of Education. "As … districts heard about the curricular and instructional resources available to them (through the STEM Council), I saw a number of districts interested in offering STEM courses or incorporating STEM content and practices into their current curriculum," Kilibarda recalled in the pre-endorsement era.[2] "In many of those districts, teachers were excited about offering STEM experiences but were unsure what that entailed or did not feel adequately prepared in one or more of the STEM content areas." That was her motivation to lead developments. Kilibarda was guided by a deep commitment to integration across *S*, *T*, *E*, and *M*, yet "Few educators had the experience to design coherent units of instruction that were

---

2 Quotes in this section are from the author's personal communications with Kilibarda in July 2016.

driven by authentic phenomena or problems. Even fewer K–8 educators had experience with engineering and computer science—two essential areas of STEM." Coincidentally, districts were moving toward specialists at the upper elementary who could teach science and math, while middle schools increasingly sought multisubject expertise. It was prime time for a new endorsement that would prepare interdisciplinary thought leaders for the state's rapidly evolving STEM climate.

During the design phase for Iowa's STEM endorsements, Kilibarda and her team navigated a similar challenge as did Edgington—drawing from an eclectic committee a singular definition for STEM and what a person with a STEM endorsement would and would not be allowed to teach. Central to the discussion was the question of appropriate content and peda-

gogical knowledge in each of the STEM subjects. "For example in science," Kilibarda illustrated, "we had to wrestle with whether or not a requirement of coursework in all of the science domains (life, physical, Earth) would be required or if in-depth knowledge in one domain (i.e., physical) would be sufficient." And in the content areas of computer science and engineering, they grappled with types and amounts of courses most suitable. Would existent courses in those fields even work for their needs? The committee unanimously considered an authentic STEM experience essential to the endorsement, but struggled with parameters and monitoring of research experiences, internships, and other variants. They leaned heavily on

A Framework for K–12 Science Education (NRC 2012) as well as the *Iowa STEM Education Roadmap* (Governor's STEM Advisory Council 2011) and existent interdisciplinary teaching endorsements in Iowa (the middle school math-science endorsement) and other states to inform the model.

What distinguishes Iowa's STEM endorsees from their traditional math- and science-endorsed peers is the amount of coursework, 12 credits each, across fields, including engineering. They also take teaching methods courses that cross disciplines. But in Kilibarda's opinion the greatest demarcations are the 30-hour STEM field experience, and the unique set of components to the STEM methods/pedagogy courses that are not included in standard math or science methods courses. Those components are in **bold type** under *The Iowa Board of Educational Examiners Teaching Endorsement #975 for K–8 STEM* section that follows.

CHAPTER 7

Reactions to the new STEM endorsements vary across stakeholder groups. "It appears administrators are more interested in having the STEM endorsement as an option for their current inservice teachers to obtain rather than one they would look for in a new hire," Kilibarda discovered. Elementary administrators are keen on reading and special needs endorsements in the current milieu. Teachers "express interest but they are still a bit concerned to see if this will be a marketable endorsement." Preparatory institutions across the state are at varying stages of developing one or more of the STEM endorsements. Although adding a new program and redesigning courses can take considerable time at the college/university level, "a number of institutions have put courses and programs of study into place," Kilibarda notes. One of them, at Drake University, is profiled shortly.

Three years into the Iowa STEM endorsements, 4 colleges were offering them, with 11 more in various stages of development. Kilibarda believes the next challenge is for excellent engineering education coursework to be developed that provides appropriate content for K–8 teachers. "In particular, many of our private colleges and universities do not have engineering departments so it is difficult for them to offer the appropriate engineering coursework," she says. The computer science (CS) component needs to be ironed out as well: Will CS courses have "a focus only on coding or on computational thinking or on programming logic?" she asks. And, bringing a consistent standard to the 30-hour STEM experience will become more and more important. The design committee's vision was that "endorsement candidates must be involved in a significant research experience or an internship or externship that allows them to engage in the work of the STEM business/industry/nonprofit." The need for guidelines and templates is emerging. Ultimately, the success of Iowa's STEM endorsements come down to market value: "Perhaps narratives highlighting the classroom innovation and instructional successes of educators with the STEM endorsement would help administrators envision the instructional leadership potential of educators with the STEM endorsement," says Kilibarda.

The Iowa Board of Educational Examiners Teaching Endorsement #975 for K–8 STEM

Iowa's STEM teaching endorsements may be accessed alphabetically at the Board of Education Examiners website (*www.boee.iowa.gov/endorsements/endorsements_teacher_gened.html*). The K–8 endorsement requirements follow:

(1) Authorization. The holder of this endorsement is authorized to teach science, mathematics, and integrated STEM courses in kindergarten through grade 8.

(2) Program requirements. Be the holder of the teacher–elementary classroom endorsement.

(3) Content.

- Completion of a minimum of 12 semester hours of college-level science.

- Completion of a minimum of 12 semester hours of college-level math (or the completion of Calculus I) to include coursework in computer programming.

- Completion of a minimum of 3 semester hours of coursework in content or pedagogy of engineering and technological design that includes engineering design processes or programming logic and problem-solving models and that may be met through either of the following:

  - Engineering and technological design courses for education majors;

  - Technology or engineering content coursework.

- Completion of a minimum of 6 semester hours of required coursework in STEM curriculum and methods to include the following essential concepts and skills:

  - **Comparing and contrasting the nature and goals of each of the STEM disciplines;**

  - **Promoting learning through purposeful, authentic, real-world connections;**

  - **Integration of content and context of each of the STEM disciplines;**

  - **Interdisciplinary/transdisciplinary approaches to teaching (including but not limited to problem-based learning and project-based learning);**

  - **Curriculum and standards mapping;**

  - **Engaging subject-matter experts (including but not limited to colleagues, parents, higher education faculty/students, business partners, and informal education agencies) in STEM experiences in and out of the classroom;**

  - **Assessment of integrative learning approaches;**

  - **Information literacy skills in STEM;**

  - **Processes of science and scientific inquiry;**

  - **Mathematical problem-solving models;**

  - **Communicating to a variety of audiences;**

  - **Classroom management in project-based classrooms;**

  - **Instructional strategies for the inclusive classroom;**

  - **Computational thinking;**

  - **Mathematical and technological modeling.**

- Completion of a STEM field experience of a minimum of 30 contact hours that may be met through the following:

  - Completing a STEM research experience;

  - Participating in a STEM internship at a STEM business or informal education organization; or

• Leading a STEM extracurricular activity.

Profile of a STEM Endorsement Program: Drake University

A select few institutions of higher education in Iowa were nimble enough to roll out STEM endorsement programs for their preservice teachers on the heels of the state's rule making. One of the first was Drake University in Des Moines. Dr. Jerrid Kruse, associate professor of science education and chair of the Teaching and Learning Department was Drake's STEM endorsement champion. His institution has all three endorsement pathways up and running: the K–8 STEM, the 5–8 STEM, and the K–12 STEM specialist.

"We've had more than 70 students enrolled in our STEM coursework," Kruse recounted recently, noting that the figure includes both undergrad preservice, graduate preservice, and graduate inservice teachers.[3] Time will tell how many complete a full endorsement. He believes the endorsements are of great interest among teachers-to-be and area practitioners, although impediments present a challenges: "I think the idea of STEM is extremely popular …, but the pragmatics of taking the coursework can get in the way." Drake's subject-matter coursework requirement for the endorsement is higher than what the state requires. Given the packed programs of study for preservice teachers and the limited night or weekend content course offerings for inservice teachers, "they struggle to complete the content coursework," according to Kruse.

When they first set about designing the endorsement pathways, Kruse and her team considered the inclusion of engineering and technology education aspects "the most significant addition to the math and science coursework." They relied on the *NGSS* for engineering integration waypoints and "created two new STEM courses focused on the natures of STEM and methods of engineering and technological design." Much overlap with best pedagogical practices in math and science made the genesis easy, though the content is novel. Today, almost all of Drake's science teaching candidates are taking the STEM courses, too.

Demand for Drake's STEM endorsees remains somewhat a leap of faith. Kruse believes that "middle schools are excited to get teachers with the STEM endorsement, but because there are so few STEM courses in the K–12 system, there is not a huge demand for the STEM endorsement." The most receptive market at the moment may be inservice, where the STEM endorsement coursework is "helping math and science teachers get better at what they do and add dimension to what they do in their classrooms."

Supply and Demand for STEM Teachers

The production of STEM-credentialed educators is on an incremental increase in contrast to the rapid proliferation of STEM schools. Administrators of STEM schools may come to favor STEM-trained and endorsed educators over disciplinary-focused teachers when, to Kilibarda's point, their unique skills and instructional leadership are recognized and valued. That mindset will drive demand. And it will have to overcome the inertia of policy and institutional

---

3 Quotes in this section are from the author's personal communications with Kruse in July 2016.

constraints, which dampen progress and rein in market forces. As pointed out by Edgington in Wisconsin, the vast majority of teachers and administrators support a STEM license option. And Kilibarda found in Iowa that K–8 administrators are interested in STEM endorsed teachers though reading and special needs command priority. In anonymous feedback, a middle school principal in rural Iowa summed up the constellation of factors bearing on the popularity of the STEM endorsement this way: "We want all of our educators, especially in the STEM areas, to integrate content and connect learning to their outside world, so in that respect the endorsement is a good thing." Principals, perhaps universally, find the STEM endorsement valuable in principle. In practice, however, its time has yet to come for many. "But right now a STEM-endorsed teacher would find herself or himself teaching five or six preps—a math, a couple of sciences, an intro to engineering, and a graphic design class because that's how we'd find it valuable," the middle school principal goes on to say. While at the same time, he or she recognizes the misuse: "Not the best use of their integrated training I know, but up the chain they are not ready for incoming high schoolers with transcripts full of STEM rather than customary math or science." STEM leaders have work to do in the policy realm around course crediting, prerequisites, graduation requirements, and instructor credentialing for the STEM license to take off. If and when it does, all but the most proactive preparatory institutions may be reactive.

STEM licensing and endorsement advocates do well to borrow pages from the medical school playbook. Rapidly evolving patient-care environments—such as those that are more personal, more holistic, more preventative, and more efficient—have pushed down to training programs new expectations for integrated, student-centered, applied, and practical approaches. The young "patients" in the care of educators have well-documented and research-based needs for learning in a personal, integrated, applied, and practical fashion. That push down to teacher preparatory programs is happening before our very eyes.

References

Brickhouse, N., and G. Bodner. 1992. The beginning science teacher: classroom narratives of convictions and constraints. *Journal of Research in Science Teaching* 29 (5): 471–485.

Bybee, R. 2013. *The case for STEM education.* Arlington, VA: NSTA Press.

Fu, M. Y., and J. Joung. *The Crimson.* 2015. Top Medical Schools React to Harvard's Curriculum Change. Sept 29. *www.thecrimson.com/article/2015/9/29/schools-react-medical-curriculum.*

Governor's STEM Advisory Council. 2011. *2011 Iowa STEM education roadmap.* Cedar Falls, IA: Governor's STEM Advisory Council. *www.iowastem.gov/sites/default/files/STEMEducationRoadmap2011.pdf.*

National Research Council (NRC). 2010. *Preparing teachers: Building evidence for sound policy.* Washington, DC: National Academies Press.

National Research Council (NRC). 2012. *A framework for K–12 science education: Practices, crosscutting concepts, and core ideas.* Washington, DC: NRC.

CHAPTER 8

The Professional Development of STEM Teachers

Building capacity for STEM intricately involves transforming teachers of mathematics, science, and other subjects into STEM teachers. Exemplary professional development models such as externships, innovation studios, community catalysts, teacher recognition incentives, and other instruments of change are used by meta-cogitators for producing the thinkers who can produce young STEM thinkers. Much as drone technology jumped out ahead of regulatory policy, STEM education leapt out front of the support systems that aid consistency and ensure high quality. This chapter examines the challenges inherent in helping everyone from content area secondary teachers to generalist elementary teachers implement a STEM learning mission.

Engineered Obsolescence of STEM Professional Development

The mission of teachers and parents is often said to be a sort of planned obsolescence—get children to the point where they can learn and grow on their own. Provide them foundational skills and beliefs and watch them soar. The whole idea is to not be (as) needed. Lots of jobs have such a goal. Orthodontists engineer their own obsolescence by fixing crooked teeth. Retirement planners should not be needed once they get their clients to the golden years. Red Cross workers—the angels that arrive at disaster scenes following floods and earthquakes—probably look forward to working themselves out of jobs. Refugee settlers plan for obsolescence by getting new Americans housed and acclimated.

The same could be said of STEM professional development. Across professions, from plumbers to podiatrists to physics teachers, periodic updates on new products and ideas will always be essential. But the necessity for transformative rebuilding of skills and knowledge is typically a once-in-a-professional lifetime sort of thing. Over the course of a 40-year career, a complete reboot instigated by dramatic new knowledge or tools is infrequent. The orthopedic surgeon learning arthroscopy in the 1970s, the auto mechanic learning computer diagnostics in the 1990s, the machinist learning CNC in the 2000s, the wildlife ecologist learning GPS in the 2010s, and the science teacher learning a STEM approach in 2020 represent major shifts in practice that, if too frequent, would set heads spinning and drive people to less frenetic careers. Revolutions in technological advancement or knowledge and skill precipitate a type of professional development that almost redefines the job. It is intensive and fundamental. Once

An advertisement for the venerable hand-cranked starter, which was engineered to its obsolescence in 1911 when Charles Kettering invented the electric motor self-starter.

transformed to the new way of doing or thinking, then we're back to periodic updates. In that sense, STEM professional development—the paradigm reconstructing type—is destined, indeed planned-for, obsolescence. A time will come when every forester has either acquired GPS skills or retired, with those coming up the professional pipeline trained in GPS out of the gates. All lathes and mills will be computer controlled by machinists whose pre-professional training set them up on the front edge. Correspondingly, educators in the

STEM arena will have been trained to implement STEM with those in succession having been well schooled in STEM preparatory programs (the subject of Chapter 7).

Who are the planners who engineer obsolescence? Futurists and innovators who know a better way and can make it happen. They bring the next big thing and render the current thing obsolete. Just such a person was John Runyon who invented CNC at MIT in the 1950s. For arthroscopy, the futurist was Japanese doctor Masaki Watanabe in the 1950s. South African scientist Peter Ryan first applied GPS to tracking penguins in early 2000s. In STEM education, a number of visionaries seemingly foresee trends, one of whom is Jan Morrison. As president and CEO of the Teaching Institute for Excellence in STEM (TIES), "the country's foremost innovator in STEM School design, STEM curriculum, and STEM instructional support to schools" (*www.tiesteach.org/about*), Morrison's knowledge of the U.S. STEM landscape is second to none, from this author's experience.

Here's what Morrison says about the obsolescence of STEM professional development: "I think that this is the time to leave professional development behind. It is an old construct."[1] Now, before rushing out to dismantle vast professional development infrastructures and livelihoods, hear Morrison out. "I think that it is time to design new systems that support from the ground up and respond to the needs of teachers and educators as well as administrators and influencers." In keeping with a recurrent theme to this point, a recast in how we perceive professional development in STEM is in order. "Education and specifically STEM education is the responsibility of the community, not just the school and its personnel." Morrison's thinking invokes the community backdrop of Chapter 3. But the current system presents constraints that serve only to invigorate STEM-minded thinkers to "use our own STEM smarts to enable much more authentic support and with more pervasive and smart touch to all within the students' own communities."

Presiding over one of the nation's esteemed STEM supporting organizations, Morrison knows as well as anyone that the cupboard is coming to be well-stocked with community-linked professional development options. But echoing an earlier cautionary note regarding the proliferation of STEM curriculum, STEM credentials, and STEM schools, the STEM professional development landscape is wildly variable.

The State of STEM Professional Development

The good news according to Morrison is that "there is high demand for top-quality STEM professional development," but the bad news as she sees it is a lack of "real understanding of what that looks like." Her opinion enjoys the backing of the National Academies. In 2014, the Committee on Integrated STEM Education of the National Academy of Sciences released a report *STEM Integration in K–12 Education: Status, Prospects, and an Agenda for Research*, which concluded as much: "Because integrated STEM education is a relatively recent phenomenon, little is known from research about how best to support the development of educator expertise in this domain specifically" (Honey, Pearson,

---

1 Unless otherwise noted, quotes in this section and the next are from the author's personal communications with Morrison in July 2016.

and Schweingruber 2014). Well-meaning service providers have gotten out ahead of the research, it would seem, from this scathing status assessment from the National Academies of Science, Engineering, and Medicine report of the Committee on Strengthening Science Education Through a Teacher Learning Continuum in 2015 called *Science Teacher's Learning: Enhancing Opportunities, Creating Supportive Contexts*, "There has been a recent proliferation of external vendors, funders, and providers of professional development in science. … The quality of these services and providers is highly variable. This variability promises to increase as vendors sell materials and services that are aligned only superficially to the *NGSS*. As the field grows increasingly crowded, it becomes more difficult for system leaders to identify high quality resources that will offer the kinds of support science teachers need in this age of reform" (Wilson, Schweingruber, and Nielsen 2015).

STEM educators love a challenge, and professional development presents a ripe opportunity. Prevailing tactics often fall short of ambitions, concluded The New Teacher Project (TNTP) in its 2015 report *The Mirage: Confronting the Hard Truth About Our Quest for Teacher Development*. After a deep analysis of professional development at three large,

typical American school districts, it estimated that about $18,000 per teacher per year is invested in faculty training. That equates to roughly $8 billion per year spent by the country's 50 largest school districts combined, wagered on enrichment programs to improve teaching. But as the report's feisty title betrays, bang for the buck is more of a pop: Nearly 70% of teachers' performance measures remained constant or declined over the three-year study. Improvement tied to any particular professional development experience was elusive, and ephemeral (TNTP 2015). The organization 100kin10, introduced in Chapter 5 as an umbrella user of the phrase *STEM teacher* and revisited in Chapter 7 for its "Grand Challenges" to STEM excellence, included STEM professional development as one of those challenges. Paraphrasing from its post at *https://100kin10. org/approach*: Too often, professional development does not satisfy STEM teachers' professional learning and growth needs because STEM content is not integrated, training to deliver active learning experiences for students is inadequate, and (reiterating Morrison's critique) the profession lacks a clear understanding of the roles, goals, and effects of various professional development experiences.

All of this is to say that the current state of STEM professional development presents a golden opportunity for innovative, entrepreneurial experts to establish a bar of excellence aligned to the best practices in STEM education. The recipe is well known. And it has become a matter of national import, considering the U.S. Congress embedded the following STEM mandate into its 2015 Every Student Succeeds Act (ESSA):

> *STEM Master Teacher Corps (ESEA, as amended by ESSA, title II, section 2245): From funds reserved for title II national activities, grants may be awarded to: (1) SEAs to enable them to support the development of a statewide STEM master teacher corps or (2) SEAs or nonprofit organizations in partnership with SEAs to support the implementation, replication or expansion of effective STEM professional development programs in schools across the State through collaboration with school administrators, principals, and STEM educators.*

Best Practices in STEM Professional Development

Brenda Wojnowski and Celestine Pea buoyed the stormy seas of STEM professional development with their 2013 treatise *Models and Approaches to STEM Professional Development*. From the colorful history of professional development in science education to the present and future realization of the promise of the *Next Generation Science Standards* (*NGSS*), the authors and their contributors identify the critical aspects shared by successful models. "Professional development that integrates content learning with analysis of student learning and implications for instruction can impact student learning," the authors derive from research into best practices (Wojnowski and Pea 2013). Critical aspects have been a focal point of researchers to address the oft-cited shortcomings identified in the previous section.

"A most basic premise of all quality STEM professional development programs …" stated the Committee on Integrated STEM in its National Academy of Sciences report, queue drum roll … is that "if teachers have not themselves experienced integration of science, mathematics, technology and/or engineering, they are not likely to teach integrated curricula for these subjects in their classrooms" (Honey, Pearson, and Schweingruber 2014). So there you have it, litmus test #1 for gauging the authenticity of STEM professional development. Following on, the Committee on Strengthening Science Education Through a Teacher Learning Continuum reported their version of critical aspects of STEM professional development as the following (Wilson, Schweingruber, and Nielsen 2015):

- Content focus—learning opportunities for teachers that focus on subject matter content and how students learn that content (in an integrated fashion)

- Active learning—for example, observations of expert teachers, interactive feedback and discussions, reviews of student work, leading study groups, and connecting with the community

- Coherence—consistency with other learning experiences and with school, district, and state policy

- Sufficient duration—both the total number of hours and the span of time over which the hours take place

- Collective participation—ideally teams of teachers (and administrators, community members) from the same school, grade, or department

Reflecting to this point on the tenets of STEM teaching and STEM schools outlined by the University of Chicago's STEM schools study and the Opportunity Structures for Preparation and Inspiration report discussed in Chapter 5 (LaForce et al. 2016), STEM professional development ought to additionally equip teachers to do the following:

- Link classroom learning to career pathways.

- Develop curriculum (and assessments) around local issues and problems.

- Distribute leadership and share decision making with learners in a safe, failure-tolerant environment.

- Successfully support students underrepresented in STEM.

The rest is mere shopping around. It is beyond the scope of this chapter to identify all the wonderful programs available to STEM educators that fulfill most and sometimes all of these key components of professional development. Instead, the next two sections highlight widely respected delivery organizations, and locally cultivated exemplars, respectively.

STEM Professional Development Resources

The **STEM Learning Ecosystems Initiative** is designed and supported by the STEM Funders Network, which is managed by Jan Morrison's TIES. The mission is essentially to provide resources for local and regional consortia to assemble, organize, and thereby advance practices that promote active, inquiry-based learning to (1) build students' competence and self-efficacy in STEM; (2) deepen their understanding of their current and future potential to solve complex problems; and (3) strengthen their social-emotional skills, including persistence, resiliency, creativity, problem solving, and collaboration. A great variety of vetted tools and models of professional development are provided. (For more information on this program, go to *http://stemecosystems.org.*)

NSTA Curators are a competitively selected "cadre of educators from around the country to establish a library of vetted resources to help teachers make the instructional shifts in their classrooms that the *NGSS* call for. The curators, who receive regular training and use the EQuIP rubric to evaluate resources, offer exemplars of both the kinds of resources educators now need and the process by which educators can take existing materials and adjust them to support *NGSS* instruction." (For more information on this program, go to *http://ngss.nsta.org/about-curators.aspx.*)

As to **EQuIP (Educators Evaluating the Quality of Instructional Products),** NSTA and Achieve developed a rubric system "by which to measure the alignment and overall quality of lessons and units with respect to the *Next Generation Science Standards.*" The intent of the project is ultimately to provide science teachers with exemplars/models to use within and across states with complete confidence. (For more information on this program, go to *www.nextgenscience.org/resources/equip-rubric-lessons-units-science.*)

The **NSTA Learning Center** is a self-directed cornucopia of resources, including research digests and activity pools, events such as webinars and online conferences, and connections including topical forums. And "e-PD resources" are indexed to individual needs, based on a self-prescribed learning plan complete with certificates. (For more information on this program, go to *http://learningcenter.nsta.org.*)

The **National Education Association STEM Resources** is a recently launched compendium of resources including grants, articles, classroom activities, and links to professional development resources at PBS (*www.pbs.org/teachers/stem*) and NASA's electronic professional development network (*www.nasa.gov/audience/foreducators/index.html*). For more information on this program, go to *www.nea.org/home/stem.html.*

International Technology and Engineering Education Association STEM Professional Development is an online "STEM+ Center for Teaching and Learning" that supports educators "through face-to-face professional development, webinars, and an online learning community that prepare educators to be Integrative STEM professionals. Professional development opportunities include collaborative learning communities, summer institutes, and on-site workshops." For more information on this program, go to *www.iteea.org/STEMCenter/EbD-PD.aspx.*

The **National Alliance for Partnerships in Equity (NAPE)** is a global leader in the STEM education equity professional development space, providing "a variety of professional development opportunities to school administrators, middle school, high school, and community college educators, and career counselors." NAPE's professional development focuses on improving underserved populations' access to and success in educational and training programs that lead to high-skill, high-wage, and high-demand careers. Such populations include students pursuing occupations that are nontraditional for their gender, underrepresented groups (including women and girls) in science, technology, engineering, and mathematics (STEM), and special populations (English-language learners, students with disabilities, single parents and pregnant women, displaced homemakers, and economically disadvantaged students). For more information on this program, go to *www.napequity.org/professional-development.*

Discovery Education offers the following three levels of professional development, each having earned the New York Academy of Science's Global STEM Alliance certification described in Chapter 6. For more information, visit *www.discoveryeducation.com/what-we-offer/stem/index.cfm:*

1. STEMformation takes three years to STEM-certify an administrator, master teacher, or teacher on lesson and unit design as well as vertical alignment of STEM units.

2. STEM Leader Corps takes four years to produce STEM champions at the school and district level through training in shared leadership, transdisciplinary curriculum, and STEM instructional progression through job embedded coaching and administrative leadership support.

3. STEM Foundations professional development is an array of discrete offerings of short duration ranging from STEM lesson development to community engagement strategies and more.

Homegrown STEM Professional Development

Chapter 4 declared that locally developed partnerships, curriculum, and assessments are the best-case scenario for building STEM community connections, although vendors and developers often helpfully provide scaffolds for homegrowing connections. *Edupreneurs* design and launch those innovative learning solutions. There are plenty of edupreneurs at work in the STEM professional development space as well. Examples follow.

APEX and BIG

Two of the nation's leading edge, innovative high school models are the APEX (Aspiring Professional Experience) program of the Waukee Community Schools in Waukee, Iowa, and the BIG learning partnership of Cedar Rapids and College Community school districts in Iowa. Both were designed by edupreneurs with the goal of producing young entrepreneurs in the STEM space. How they each go about professionally developing their faculty is a lesson in culture shift.

APEX teachers take part in not only the school district's standard professional development but also the "professional development delivered through our business partners," said Michelle Hill, director of the Waukee APEX.[2] The curriculum is said to be "passion-driven and personalized" which calls for unique training to learn a particular skill necessary on the business end, often on a just-in-time basis. "An example of this might be particular

2 Quotes in this section are from the author's personal communications with Hill and Miller in July 2016.

software, like Trello or AutoCad, or methodology, like Agile," Hill recounted, speaking of recent professional development needs delivered by business partners. Easterly by 100 miles, the BIG teachers engage daily with professionals from business, nonprofits, and government, according to Troy Miller, director of strategic partnerships. Since students' projects are intensely public and community-linked, teachers' "depth of development compounds daily as they each experience the most relevant management techniques, technologies, and societal movements as experienced through serving as an Instructor for students' public-facing projects." At both APEX and BIG, professional development focuses on "a broad understanding of all the elements that cause communities to thrive" as Miller put it. Hill highlighted that APEX has a reciprocal relationship with district partners who have found that district-offered professional development helped provide them a "new perspective on training topics such as mentoring," which "were valuable for their staff to use while working with our students and their entry-level employees."

Both BIG and APEX embrace Morrison's vision for a community-based professional development enterprise. Miller firmly believes that "no entity has more relevant and timely curriculum sourcing for students than the living, breathing community. Public education needs to invite the public to re-engage." Westward to APEX, the community, including administrators and parents, join in on the professional development given such timely topics as design thinking, mentoring methods, Agile methodologies, growth mindset, and other sometimes more technical training. The future of professional development at both APEX and BIG will always be tied to the needs of the economic sector, primarily skills related to high-demand careers. In that way, both models have essentially flipped the script: Schools are functioning as economic development engines and business partners are contributing to educational innovation.

Iowa Teacher Externships

Teacher externships were highlighted in Chapter 4 as outstanding entry and sustenance models of school–business partnerships. "When am I ever gonna need to know this?" was the common lament of students that inspired Iowa's externship program. It turns out that teachers in the applied workplace benefit mightily in terms of professional development as well. After six weeks at a hospital doing inventory and efficiency studies, a mathematics teacher comes away with visceral understandings not only of what 21st-century skills look like in practice, but also how nurses and CFOs use numbers to maintain operations. After a summer-long immersion at a paint manufacturer, a chemistry teacher heads back to school with a whole new appreciation for collaborative skills, communication, attention to detail, and commerce, to name a few attributes to be more heavily emphasized in class (let alone the chemistry content update, which many externs consider valuable, but a distant second to workplace enlightenments).

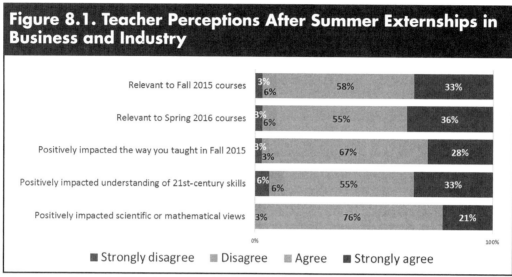

Figure 8.1. Teacher Perceptions After Summer Externships in Business and Industry

Source: Iowa Governor's STEM Advisory Council; reprinted with permission.

A program initially funded by the National Science Foundation (grant number DRL1031784), Iowa's externships program has been studied to the hilt by the Center for Social and Behavioral Research at the University of Northern Iowa. Teacher benefits align closely to the STEM teacher skillsets *a–g* discussed earlier, including the following (Figure 8.1):

- Ability to career-coach students, especially regarding skilled trades
- Ability to apply course content outside of the classroom
- Ability to operate collaborative, project-based learning
- Ability to convey 21st-century skills

Evaluators determined that 90% of participants in summer externships consider the experience their most valuable professional development. These and the following findings of the evaluators are summarized as Chapter 12 in the 2012 NSTA Press publication *Exemplary Science for Building Interest in STEM Careers,* edited by Robert E. Yager.

Workplace hosts who take in teacher-externs for summer also derive significant benefits, including the following (Figure 8.2):

- Teachers make genuine contributions to operations.
- Insights into school culture and where to best "make entry" are established.
- Long-term school–business partnerships result.
- Tens of thousands of dollars in added value is gained from the presence of teachers.

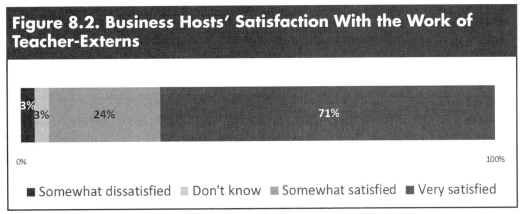

Figure 8.2. Business Hosts' Satisfaction With the Work of Teacher-Externs

3%

3%

24%

71%

0%

100%

■ Somewhat dissatisfied ■ Don't know ■ Somewhat satisfied ■ Very satisfied

Source: Iowa Governor's STEM Advisory Council; reprinted with permission.

Students of externs show little gain compared to peers in terms of standardized test scores in mathematics or science, though interest gains in science, mathematics, and STEM careers increase. For female students, the difference was statistically significant in 2015 (Figure 8.3), although evaluators appropriately offer the caveat that these results are associational and could be affected by other factors not measured as part of the evaluation (Pollock and Losch 2015).

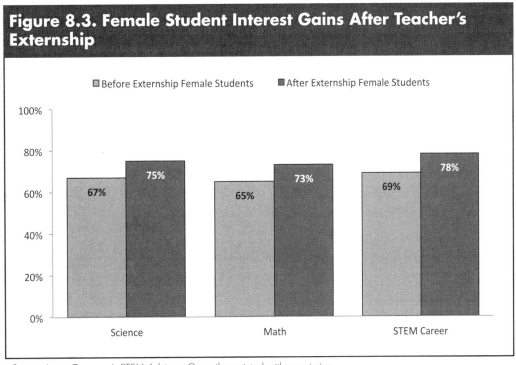

Figure 8.3. Female Student Interest Gains After Teacher's Externship

■ Before Externship Female Students   ■ After Externship Female Students

	Science	Math	STEM Career
Before	67%	65%	69%
After	75%	73%	78%

Source: Iowa Governor's STEM Advisory Council; reprinted with permission.

CHAPTER 8

STEM: A New Construct for Professional Development

STEM professional development—the sea-change redefinition kind that equips a science or mathematics teacher to enliven a STEM classroom—may be a practice in engineered obsolescence. When all willing educators have been *STEM-ed* and freshly minted professionals are emerging from higher education institutions well-versed in STEM practices, a different sort of professional development—a community focused partnership cultivation—will (and does

Photo courtesy of the author.

Taaf Vermeulen, technology education teacher at Albia Community High School in Albia, Iowa, working with the company Musco Lighting.

at APEX, BIG, and through externships) prevail. Until then, a potpourri of options exists for helping content-specific teachers metamorphose and hone their STEM skills.

References

Honey, M., G. Pearson, and H. Schweingruber, eds. 2014. *STEM integration in K–12 education: Status, prospects, and an agenda for research.* Washington, DC: National Academies Press.

LaForce, M., E. Noble, H. King, J. Century, C. Blackwell, S. Holt, A. Ibrahim, S. and Loo. 2016. The eight essential elements of inclusive STEM high schools. *International Journal of STEM Education.* doi 10.1186/s40594-016-0054-z.

Pollock, N., and M. Losch. 2015. Real world externships for teachers of math and science, 2014–2015. Report of findings, Center for Social and Behavioral Research, University of Northern Iowa, Cedar Falls.

The New Teacher Project (TNTP). 2015. *The mirage: Confronting the hard truth about our quest for teacher development.* New York: The New Teacher Project. *http://tntp.org/publications/view/the-mirage-confronting-the-truth-about-our-quest-for-teacher-development.*

Wilson, S., H. Schweingruber, and N. Nielsen, eds. 2015. *Science teacher's learning: Enhancing opportunities, creating supportive contexts.* Washington, DC: National Academies Press.

Wojnowski, B., and C. Pea. 2013. *Models and approaches to STEM professional development.* Arlington, VA: National Science Teachers Association.

Yager, R. E., ed. 2012. *Exemplary science for building interest in STEM careers.* Arlington, VA: NSTA Press.

CHAPTER 9

Hopes, Hazards, and Horizons

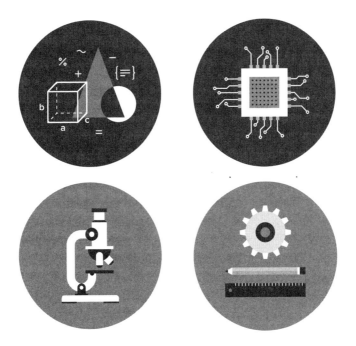

The STEM community is responsible for the most invigorating influencer of schooling since the advent of public education. *Hope* rides on STEM to bridge education to the broader community. *Hazards* may lurk in policy inhibition and comfort in constancy. Yet, the *horizon* is replete with opportunity to provide to all learners a useful education. The last word on these matters has been reserved for leading thinkers, both producers and consumers across the STEM spectrum, who weigh in on where we go from here. Classroom practitioners, industry advocates, community catalysts, parents, elected officials, and the students for whom all toil, each offers unique threads of thought that together form a fabric of the future of STEM.

CHAPTER 9

About This Chapter

Seventeen contributors responded to these two questions:

1. What is the one thing about America's STEM movement that most *worries* you that we need to get right to fulfill the promise of STEM?

2. What is the one thing about America's STEM movement that most *wows* you about the great promise of STEM?

Respondents appear in the order shown below (alphabetical and alternating genders; some titles and affiliations have changed since the time of this writing):

1. Steven Barbato, Executive Director and CEO, ITEEA (International Technology and Engineering Educators Association)

2. Edie Fraser, CEO, STEMconnector and Million Women Mentors

3. Vince M. Bertram, EdD, President and CEO, Project Lead The Way

4. Riley Hubbart, junior at Clinton High School in Iowa pondering a college major in biomedical engineering

5. James Brown, Executive Director, STEM Education Coalition

6. Mimi Lufkin, CEO, National Alliance for Partnerships in Equity

7. Rodger W. Bybee, PhD, educational consultant and author of *The Case for STEM Education: Challenges and Opportunities* (2013)

8. Linda P. Rosen, PhD, CEO, Change the Equation

9. David Etzwiler, CEO, Siemens Foundation

10. Yvonne M. Spicer, EdD, DTE, Vice President, Advocacy and Educational Partnerships, National Center for Technological Literacy, Museum of Science, Boston

11. Quentin Hart, Mayor of Waterloo, Iowa

12. Mary Wagner, PhD, Senior Vice President, Global Product Innovation and Chief Food Safety and Quality Officer, Starbucks Coffee Company

13. Kwizera Imani, sophomore majoring in aerospace engineering at Iowa State University and member of Iowa Governor's STEM Advisory Council

14. Isa Kaftal Zimmerman, EdD, distinguished educational consultant and author who served on the Massachusetts and Iowa Governors' STEM Advisory Councils

15. Robert E. Yager, PhD, Professor Emeritus of Science Education, University of Iowa

16. Jennifer Zinth, Director of High School and STEM Programs, Education Commission of the States

17. Kim Reynolds, Governor of Iowa (Alphabetical protocol abandoned, the final word of this book is reserved for Governor Reynolds, who co-chairs the Governor's STEM Advisory Council.)

Steven Barbato

Executive Director and CEO, ITEEA (International Technology and Engineering Educators Association)

The one thing that most worries me most about America's STEM movement that we need to get right is the opportunity for ALL STUDENTS (not just the best and brightest) to become STEM literate. By this, I am referring to their ability to not only be able to know and understand basic science and mathematical facts and concepts as they apply to the natural and known world, but to also have the capacity to observe real-world situations and apply their STEM knowledge in a way that embraces and employs the technological problem-solving and engineering design processes to effectively create viable solutions.

Designing STEM-based solutions under constraints to address complex human needs, challenges, and wants is a "must-have" skill for all citizens in the 21st century. My concern is the reality that most American students are being taught STEM in a "silo" delivery system within each discreet content area, never having the opportunities to integrate their knowledge through real-world scenarios. The decreasing number of qualified STEM teachers to teach truly integrative STEM topics (see Love, Love, and Love 2016) coupled with the decreasing number of programs that produce them, increases the challenge of providing enough authentic experiences for all children to become STEM literate. We must provide a broad range of "Integrative STEM Education" learning experiences at every grade level, providing all students with multiple opportunities to explore, create, and innovate by using their STEMcentric skillset!

The one thing about America's STEM movement that most wows me about the great promise of STEM is the enormous groundswell of interest we are seeing from the abilities of children of all ages! It is incredible to see how truly Integrative STEM Education taps into the innate abilities that all students possess to problem solve, create solutions, and experience success—especially when success may be first disguised as failure! There are countless examples of how invention and innovation occurs during and after many unsuccessful attempts to solve a problem (e.g., the light bulb by Thomas Edison and Post-it notes by 3M). We have the ability to influence the productiveness and success of our

Continued

country and the world by creating highly capable problem-solving young adults who are ready to address the 14 grand challenges outlined by the National Academy of Engineering. Integrative STEM Education builds STEM-capable students to create solutions that are achievable and sustainable to help people and the planet thrive.

To Barbato's skillset "must-have" for all future citizens, Edie Fraser adds another must-have—leadership cultivation.

Edie Fraser
CEO, STEMconnector and Million Women Mentors

What worries us is the teachers, their skill sets and their capacity to inspire, to instill excitement about STEM careers. We at STEMconnector are inspired every day by models at the local level, by programs that are being scaled up, and know impact can truly be great. It can only happen if we build leadership at every level of our system. It is our joy to work with the STEM program in Iowa, with a great lieutenant governor who is championing best programs, stellar education, and the pathway to Great STEM jobs and yes, those that pay and know Iowa and every other state and community can experience progress if the entire system propels the students into great careers. Go Iowa!

What gives me hope? So much. We see so many examples of great programs that we champion each day at STEMconnector. We note programs that may start small, but then are built to reach millions. Leaders are everywhere and we leverage their support. We call in the private sector *CEOs*, and all leaders to step up and build private-public partnerships as never before in this nation. It's beginning to work as we realize that only through partnerships and leadership can our communities, states and nation compete for being labeled "great" in STEM jobs and rewarding STEM careers. Skill up and recharge ensure that nationally, not only in Iowa but in every state in the community, we can show STEM success. We aim to take education to HIRE education.

Connected partners are Fraser's passion, whereas Vince M. Bertram zooms in on equity, especially early in the lives of our youth as the great hope and challenge for the future.

Vince M. Bertram, EdD
President and CEO, Project Lead The Way

With STEM education, our greatest challenge is also our greatest opportunity. It's a challenge we face in all education, but is most pronounced in the STEM disciplines: equity. We have a moral and economic imperative to provide all students with access to high-quality STEM education opportunities. All students—from those in our country's urban centers to rural areas, white and black, rich and poor—need access to transformative learning experiences that will prepare them to thrive in our rapidly advancing world.

The 2016 *U.S. News & World Report*/Raytheon STEM Index highlighted the significant gaps between men and women, and whites and minorities, in the STEM fields. According to the report, "As the number of white students who earned STEM degrees grew 15 percent in the last five years, the number of black students fell by roughly the same margin." Women's interest also decreased slightly from 2015 to 2016 (Neuhauser and Cook 2016). Similarly, the National Action Council for Minorities in Engineering notes that while members of racial minority groups "constitute 30 percent of the U.S. population … they earned only 12.5 percent of all [bachelor's degrees in engineering] in 2011." (National Action Council for Minorities in Engineering 2017) These statistics paint a bleak picture of our nation's educational landscape. Yet equity in STEM education is also where I see our greatest promise. We can shift the tide of disparaging inequity plaguing our education system and ensure that all students develop the knowledge and in-demand skills to take advantage of the high-tech, high-paying jobs of the future. We will make strides through a more concentrated and intentional effort at the local, state, and perhaps even federal level.

The issue of educational equity we face today is not unlike the inequities in athletics before the passing of Title IX, the 1972 law that helped level the playing field for girls and women in high school and collegiate athletics. Between 1972 and 2013, the number of women participating in high school varsity athletics grew from 295,000 to more than 3.2 million. Moreover, between 1972 and 2015, the number of women competing in intercollegiate athletics grew from 30,000 to more than 200,000. Title IX disproved the myth that female students simply weren't interested in sports. I firmly believe that the same opportunity exists with underrepresented students in the STEM fields. We must break down the barriers in education and provide these students opportunities in STEM, rather than dismissing it as something they either aren't interested in or as disciplines that are too "tough" for them.

Continued

And it is not just enough to provide the opportunity, we must reach these under-represented populations at a younger age. The earlier students are exposed to the STEM fields, the more comfortable they'll be with them, and as a result, the less they'll be intimidated by them.

We also have to continue breaking down the stereotypes—both racial and gender—that exist in many STEM disciplines that perpetuate pervasively low expectations for too many students. We need to provide these students with the means to develop the skills, knowledge and confidence to pursue STEM subjects and careers. Doing so will put them on a path to educational and career success, strengthening our nation's education system, industries, economy, and security.

Title IX-style crusade to elevate the expectations of, and opportunities for, underrepresented students in STEM is Bertram's widely shared priority. Student Riley Hubbart trains her eye squarely on teacher preparedness beyond accommodating diversity, to walk the STEM talk in terms of technology integration.

Riley Hubbart

Junior at Clinton High School in Iowa pondering a college major in biomedical engineering

I am concerned STEM is a passing fad in education. Every other year it seems there is some new method or a new focus. Teachers are inefficiently teaching. By the time teachers adapt to the new method, there's another method that administration implements. The varied emphasis on technology seems to be slowing down education more than speeding it up. In my experience, the incorporation of computers is strained and feels unnatural. Teachers are rarely knowledgeable in regards to technology and are simply trying to appease their superiors with the use of computers in their students' projects. Teachers are often unwillingly forced to participate in this new trend, and many of them are resistant. Another problem is the uneven ability level of students when it comes to computers. Some students have access to computers and learn the material. Contrarily, others have limited access and spend most of their time trying to use the computer instead of learning the material. In theory, incorporating more technology into teaching is a good idea. In our everyday lives, technology has been put to great use. The most obvious example is our cell phones, which many of us cannot imagine living without. In the classroom, however, the incorporation has not been smooth. Teachers are spending too much time learning new "research-backed methods" at the expense of the students' education.

Continued

> The most promising thing about STEM is its modern approach. As a modern education movement, it coincides with modern social movements, such as equality for women and minorities. Schools work to include more girls and ethnic minorities in STEM courses and academic clubs. In terms of modern teaching, STEM encourages a data-driven approach to determine the best methods of teaching. One approach I enjoy stems from how educators are starting to realize that just about anything can be looked up on the internet. Teaching is now more about showing students *how* to learn rather than raw material and memorization. STEM takes a modern approach to improving the education problem in our country. STEM focuses on subjects that will make our country competitive globally, and recognizes that diversity is necessary to accomplish this.

Whereas Hubbart sees risk in STEM as a passing fad if schools underadapt to technology, James Brown identifies the risk of inconsistency inherent in distributed resource allocation and national leadership.

James Brown
Executive Director, STEM Education Coalition

> The dramatic increase in state and local autonomy enacted by the Every Student Succeeds Act in 2015, is going to be a double-edged thing for the STEM movement. In many places, states and districts will use the new authorities and new federal funding available to advance STEM education goals more aggressively. But in other places, I think we will see some backsliding—and that troubles me greatly. The most worrisome part is that we won't know exactly how or where these two different trends are occurring until at least 2017, possibly longer, or which trend will dominate.
>
> I also think we need more high-level policy-maker champions for STEM issues at the national level, as well as in states. Five or 10 years ago, when the STEM movement was still pretty new, we had more champions in Congress for STEM education than we do now. We still have strong champions there, but we need more to replace some of the champions that have retired or moved into other leadership roles. The next Presidential administration is also a big question mark. President Obama spent a lot of time focusing on STEM issues, but there is so much work left to do. We are seeing a lot more visible leadership on STEM issues from governors now, which is very good.
>
> Looking into the future, I am most excited about the way in which so many STEM programs are now reaching the scale where they can really tip the balance in terms of

Continued

> outcomes. We are educating more STEM teachers now. Hundreds of thousands of students are engaged in STEM competitions every year. High-quality STEM courses are expanding into underserved schools. Hands-on learning is becoming much more common and being reinforced by new, better science standards.
>
> But the fact remains that while our best schools are doing better with regard to STEM, our struggling schools are still suffering from big-time opportunity gaps. Closing those gaps—and bringing high-quality STEM opportunities to places for the first time—is where the real fight will be in the coming years.

Burgeoning learner opportunities in STEM, albeit less blanket in coverage than patchwork across the schools of America, define the current times for Brown. That patchy opportunity gap drives Mimi Lufkin to focus on teachers as keepers of the trust whose collective influence on learners' STEM trajectories needs to be recognized and reinforced.

Mimi Lufkin
CEO, National Alliance for Partnerships in Equity

As the leader of an organization that has been focused on educational equity since the late 1970s you might expect that the one thing that most worries me about America's STEM movement is the continuing underrepresentation of women and people of color in STEM education and STEM careers. However, what worries me even more than this is the fact that there are still so many educators who either don't see this as a problem or who think that the solution is not in their control. The subtle messages that students get every day in school about whether they belong in STEM or can achieve in STEM is the most insidious form of discrimination. To fulfill the promise of STEM, every educator must understand how their implicit biases affect student engagement in STEM. This is not easy work and requires commitment on the part of everyone to dig deep, look hard, and invest in strategies that have proven to be effective. We cannot afford to approach this issue superficially as it requires a shift in culture, values, and beliefs to change.

The enthusiasm, will, and passion about STEM is unprecedented and clearly the movement has explicitly identified equity as the key to expanding opportunities in STEM. This level of awareness and the attention that the equity gap in STEM is getting nationally and globally speaks volumes to the potential to actually change the face of who is engaged in STEM. We have been successful in shrinking the gender achievement gap in STEM. We just need to help women, girls, and students of color translate that success into career choice in STEM. We have made progress in increasing women's

Continued

participation in some STEM fields: physicians were 8% female in 1970 and 37.9% in 2015 (U.S. Bureau of Labor Statistics 2017); and women earned the majority of bachelor's degrees in biology, social science, and psychology in 2014. However, we continue to see gaps in women's attainment of bachelor's degrees in engineering (14%) and computer science (18%) (Espinosa 2015); and gaps in their participation in community college manufacturing (11%) and architecture and construction (11%) programs (U.S. Department of Education 2017). Although there is still a gender pay gap in STEM, it is the smallest of all fields and women in STEM jobs earn 33% more than women in non-STEM jobs (Beede et al. 2011). There is much good news here and much work yet to be done, and I know we are up to the challenge!

We have come a long way, as Lufkin observed, though STEM is far from fulfilling its promise for all. Rodger W. Bybee suggests the headway is a credit to an effective slogan prone to underachieve unless coherent programs and policies ensue.

Rodger W. Bybee, PhD

Educational consultant and author of The Case for STEM Education: Challenges and Opportunities *(2013)*

STEM: From a Popular Slogan to Practical Policies

The acronym STEM is quite popular in American education. My concern about the STEM movement resides in the significant diversity in what STEM means when it comes to educational policies that translate to programs and practices. When asked about STEM initiatives, responses vary. The acronym may refer to activities within one of the four disciplines, a robotics competition, bringing engineers (or scientists, technicians, mathematicians) to a classroom, a constellation of problem-based activities, the recruitment of individuals for STEM careers—you get my point.

Currently, STEM is more of a slogan and less of a clear definition and plan for educational programs and teaching practices. Simply saying that STEM stands for science, technology, engineering, and mathematics only clarifies the acronym. It does not provide a definition that has implications for coherent school programs, for example. In *The Language of Education* (1960) Israel Scheffler pointed out that "slogans in education provide rallying symbols of key ideas and attitudes of education movements. They both express and foster community of spirit, attracting new adherents and providing reassurance and strength to veterans" (Scheffler 1960, p. 36).

Continued

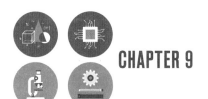

One point about educational slogans makes the transition to my positive response for America's STEM movement. The unity and allegiance to STEM can be the motivation and incentive to move beyond the slogan to meaningful and practical policies for the education system. It is important to capitalize on the slogan's contemporary support before it loses the power to rally and becomes the subject of criticism.

I have been very impressed by the potential for STEM. Local, state, and national initiatives make the claim of being STEM-oriented, with considerable support and without critical comment. This situation is the opportunity for the STEM community. The political support for STEM is broad and deep, especially from the four communities. Many 21st-century challenges—economic growth, climate change, reduction of biodiversity, vulnerabilities of the internet, energy efficiency, emerging and reemerging infectious diseases, clean water, space exploration, and healthy oceans—all depend on solutions that at least in part require interactions among science, technology, engineering, and mathematics.

The opportunity to make STEM a lasting theme in American education is now. We must ask and answer the Sisyphean question for STEM education. What should 21st-century citizens know, value, and be able to do—relative to science, technology, engineering, and mathematics? Answering this question will bring coherence to STEM and build strong and lasting policies that can translate to programs and practices in American education.

Great hope and promise reside in STEM for heading off grand challenges of our times, said Bybee, if consensus goals congeal. Otherwise, as Linda P. Rosen notes, we risk generating enthused, ephemeral activity bound to collapse for want of hard evidence.

Linda P. Rosen, PhD
CEO, Change the Equation

STEM education has become so ubiquitous that the acronym that can now stand alone without further explanation. But has the flurry of attention caused widespread enthusiasm for STEM and notable improvements in students' STEM education? I'm not so sure.

Even more troubling is the possibility that STEM *activity* has been mistaken for *progress*, which will set us up for crushing disappointment if the next few reports on student achievement, STEM degree attainment, or STEM skills shortages come up short. STEM could easily fade from the limelight and lose its resources as attention turns elsewhere.

Continued

And the postmortem analysis will suggest that enthusiasm and good intentions were insufficient.

Too many education reforms have been undone when unfocused energy or activity fails to produce sufficient results. Have we learned to be deliberate in efforts to scale up effective STEM learning? To focus on high-quality, sustained learning opportunities? To learn from others about successfully engaging young people in STEM? I have high hopes, though mixed with real worry.

Disaggregating student data is a relatively new, yet powerful, tool to strengthen education. It illuminates gaps and informs targeted, timely solutions. Before 2000, most agencies collected data of all sorts in the aggregate, lulling policy makers and parents into complacency by papering over achievement gaps and inequities in opportunities.

Thanks to disaggregation, we now know, for example, that 49% of 12th-grade students from higher income families attend a high school that offers a computer science class, while only 35% of 12th-grade students from lower income families have the same opportunity. The data are bad enough for wealthier students, but they are downright unconscionable for their lower-income peers, especially at a time when computing jobs are growing nearly twice as fast as all other jobs and pay almost twice as much.

Disaggregation also reveals that, by 2022, 9 in 10 new engineering jobs will require at least a bachelor's degree yet most of the engineering credentials African American and Hispanics earn are below the bachelor's level (U.S. Bureau of Labor Statistics 2013). The reverse is true for whites.

Data such as these have inspired educators and policy makers to devise strategies for leveling the playing field. States like Arkansas and school districts like the New York City Public Schools are requiring all schools to teach computer science classes, regardless of their students' income level. Efforts like National Math and Science's College Readiness Program are putting many more black and Latino high schoolers on a pathway to STEM careers. I am excited!

Data-driven decision making about STEM fuels Rosen's optimism for the movement to overcome Riley's fad factor and Rodger's slogan shortcoming. The Siemens Foundation, overseen by CEO David Etzwiler, leverages data on middle-skill career opportunities to heighten the country's STEM employment IQ.

David Etzwiler
CEO, Siemens Foundation

The Siemens Foundation—thanks to our support from and proximity to Siemens' businesses—understands the challenge of growing a business in an ever-shifting economy, where technologies are changing faster than ever. And while hard assets require regular investment, it's the commitment to human capital that wins or loses the day for business leaders and the nation's economy.

We often forget how drastically things have changed: Today's cars don't need a garage mechanic so much as a software engineer. And the factories of today are quiet, efficient, and highly automated requiring highly technical, computerized skill sets—often using state-of-the-art machines or robots. STEM knowledge, skills and experience are among the most important drivers of economic prosperity in the world today. Careers in STEM deliver sustained economic security for the individual, and a thriving economy through innovation.

Unfortunately, there's a stigma we continue to see—a judgment about technical jobs that require advanced learning beyond high school, but not a four year degree—an area commonly referred to as *middle skills*. In the United States, unlike other post-industrial economies around the world, too many Americans still look at these jobs and those who fill them, with unfavorable views. That's a mistake we can't afford to make considering STEM middle-skills jobs pay an average of $53,000 per year, are in high demand, and come with very little debt load. America, and its workers, need STEM middle-skill jobs.

As a company with a German heritage, Siemens' roots run deep in apprenticeships and skilled technical careers. At the core of this is an innate societal appreciation and respect for practical work.

The United States must take a broader-based approach to STEM and support the *full* pipeline of STEM needs, including practical work by skilled technicians. We need to embrace and value all levels of opportunity—not just at its highest levels. This also means encouraging those from underrepresented and underserved populations that they, too, can be part of the STEM opportunity movement.

STEM competence throughout *all levels* of our society will ensure economic success for the individual, and, importantly, will contribute to the rebuilding of a thriving middle class, and a strong foundation for the broader economic success of our nation.

The grandeur of a robust American middle class hangs in the balance for STEM, according to Etzwiler. Yvonne Spicer sees cultural relevancy as instrumental in broadening interest and participation of children of color in that vision.

Yvonne M. Spicer, EdD, DTE

Vice President, Advocacy and Educational Partnerships, National Center for Technological Literacy, Museum of Science, Boston

I'm deeply concerned about the access and opportunity gap for children of color in STEM. The recent National Assessment of Educational Progress Technology and Engineering Literacy (NAEP/TEL 2014) results illustrated that only 18% of African American students and 28% of Hispanic students are performing at or above proficient on the eighth-grade assessment. The statistical data has unveiled that many children of color are not enrolled in the prerequisite courses in high school that prepare them to pursue a career in a STEM field. Moreover, access to high-quality programs in lower grades is scarce therefore limiting exposure to what STEM is and why they should pursue it as a career pathway. For example, how inclusive are STEM programs for students of color, in low socioeconomic communities? If all students don't have access to programs, we are shortchanging our children and limiting their economic future. Over the past 11 years that I have worked at the National Center for Technological Literacy, the situation has improved in communities across the country, however overall the situation is still quite dismal. I often ask myself what would make a difference particularly for children of color. I surmise that teaching and learning about STEM needs to be culturally relevant in order to engage students of color. How often are classroom activities connected to ways in which they can improve their community or solve a problem that directly affects their lives?

The opportunity for great jobs for young people and to contribute is something that wows me. The ideas behind many of the jobs that our children will have in their lifetime, as well as the innovations that they may create, don't even exist. A reality that is somewhat daunting, but exciting simultaneously. The STEM education movement forces us to step out of our comfort zone in the classroom and critically examine practices and outcomes in education.

The greater promise is an opportunity to revamp our preK–12 education system. We have long known that our school system is broken and it certainly doesn't address the needs of all. We have an opportunity to do a much better job at addressing 21st-century skills in our schools and targeting efforts particularly to the underserved and underrepresented communities. There are school systems that have made changes, but many remain ill-prepared or complacent in the practice of rote teaching with little or no accountability for applying learning to solve existing or new challenges. I am encouraged and inspired that a new focus and direction in education is imminent.

Spicer welcomes the transformative power of STEM to revamp schooling in America. She finds good company in Quentin Hart who sees an engine of innovation cloaked in the STEM acronym, limited only by the scopes of the imaginations of STEM thinkers.

Quentin Hart
Mayor of Waterloo, Iowa

> What most concerns me is that opportunity, innovation, and education must reach and be available to everyone. From the inner city to the most rural parts of our communities, STEM access should never be limited nor should opportunities be missed. STEM thinkers must continue to think outside the box to make sure that access is never limited to a certain few.
>
> The STEM initiative is more than just a phase in how we educate our young people. It's a movement that takes the education of today and reshapes it to create the innovation of tomorrow. Innovation in transportation was what helped the Wright Brothers to be the first in flight. Innovation helped Benjamin Banneker shape historic Washington, D.C. The STEM movement will make sure that bold, new ideas, innovative sciences, and technologies will continue to thrive in the hearts and minds of those involved in its programs.

For Hart, STEM is a crystal ball into the future of innovation. For Mary Wagner, STEM is a foundational need for the nation, weakened only if we permit biases to persist rather than build systems that better engage women and girls.

Mary Wagner, PhD
Senior Vice President, Global Product Innovation and Chief Food Safety and Quality Officer, Starbucks Coffee Company

> STEM is such a critical foundational need for our country especially with an expected 2.4 million unfilled STEM jobs anticipated by 2018. Although we are making great progress on many fronts, my biggest worry is that there will still be so many women and girls left behind as a result of systemic biases in STEM education and society more broadly. Today, women are 50% of the population but represent a mere 26% of the STEM labor force (U.S. Department of Labor 2016). We must reverse this trend through targeted efforts designed to engage women and girls throughout their education and career journeys. We must continue to identify opportunities to ensure all STEM-interested women and girls have the opportunities and support structures in place to ensure success, as we advance the STEM cause in America.
>
> Through my involvement with America's STEM movement, I have been continuously moved by the great work being done across the country and continue to be

Continued

inspired by the efforts and accomplishments of others. We are making real progress throughout the United States: For example, in Washington State we have pulled together a consortium of women leaders from across several companies with the mission to ensure middle school and high school girls in Washington State are aware of and able to explore STEM careers. As a result, we now have an energized and dedicated group of women industry leaders supporting this goal. Another focus area for this group has been to foster opportunities where every postsecondary young woman has access to pursue STEM internships and jobs. I am thankful to have had the distinct privilege of partnering with these STEM leaders, as well as with several young women and girls, to craft programs that will help entice more women and girls to explore careers in STEM. I remain inspired by the depth of the passion and perseverance I have witnessed through these relationships and experiences. For this reason, I know we will succeed in America's STEM movement and I cannot wait to see our accomplishments together.

Evidence of progress in STEM inspires Wagner to optimism over the potential for full engagement of women in STEM. Collegian Kwizera Imani shares her optimism, with a similar qualifier—replicate evidence-based programs that make STEM learning more meaningful.

Kwizera Imani

Sophomore majoring in aerospace engineering at Iowa State University and member of Iowa Governor's STEM Advisory Council

As I look at the direction America's STEM movement is headed, I am pleased to say that we have a bright future. The commitment to excellence from our STEM leaders has been seen through the success of the programs that they have implemented. These programs have been discussed in the previous chapters, but I would like to acknowledge a program that is going to be critical for the future STEM students on the path of higher education.

One of the difficult things for teachers to do is keeping the students actively engaged in their learning on a daily basis. Students are best engaged in their learning when they know that material that is presented to them can be translated to real world applications. STEM has provided a solution to the teacher by introducing the Iowa STEM Teacher Externship Program. The program provides the teacher an opportunity to collaborate with local business on project-based learning. The teachers

Continued

get exposed to the latest real world application being used in the industry, and create a long-lasting partnership with the businesses. Experience such as these give a new perspective on how to approach different topics, and allow credibility to the teacher because the students know that their teacher has some experience from the industry.

My proposal to states across the nation is to introduce computer programming classes at the K–12 level that will prepare students for the college environment. Being a sophomore in the Aerospace Engineering, I am currently looking at the Engineering department and evaluating the courses that would have been helpful to the transition of the college engineering student. Having computer programing classes in the K–12 schools is a vital key to the success of future engineering students. All engineering majors are required to take a computer programming classes and most students would agree that it was difficult for them to comprehend the languages. If we were to have computer programing classes in K–12, students will develop the muscle memory and discipline of computer programing. They will begin to comprehend how to critically think through lines of code. Once this happens, a spark is ignited in their creativity and the sky is no longer the limit, for their imagination it's only the beginning.

As computer coding can open up horizons from Imani's experience, the connections inherent in STEM education (including K–16, government, and industry) control the destiny of the movement as a whole, according to Isa Kaftal Zimmerman.

Isa Kaftal Zimmerman, EdD
Distinguished educational consultant and author who served on the Massachusetts and Iowa Governors' STEM Advisory Councils

American education is replete with well-intentioned programs and practices that never achieved their promise for both understandable and regrettable reasons. STEM education cannot fall into that category. There are too many elements involved in STEM which are essential; the parts must move forward in synch, starting with students at a very early age. We need to ensure that all the parties are working together: the nation, the state, business and industry, education (K–16), and the community. Current projections of open positions and jobs, some of which are not necessarily STEM centric, but all of which require STEM education, are overwhelming for the near future, let alone the long trajectory. While political support is necessary, the movement cannot be caught up in or driven by politics. Unless the current movement

Continued

is successful in achieving STEM educated/competent citizens who also can interact constructively with the rest of the world, the nation is in a precarious position.

STEM education, of necessity, involves and engages all kinds of people: boys and girls, every ethnicity, and every business and industry. STEM education can serve as a lighthouse to bring together the variety of people we have in the United States for positive purposes: solving global challenges, protecting and improving the environment, nurturing artistic talents, creating new technologies, developing new inventions, supporting national economies, creating responsible citizens. All of these can enable individuals and families to live comfortably and healthily in an ever-more connected education system and economy. In our complex world, a STEM education is a solid base on which to learn, to live, and to understand the phenomena around us.

Zimmerman sees STEM as a unifying force of immense capacity for abating the globe's most ominous threats. She is of like mind to Robert Yager who for decades has championed a mission of "how" for education, not "what." He sees STEM as a vehicle helpful to his mission.

Robert E. Yager, PhD
Professor Emeritus of Science Education, University of Iowa

Science, Technology, Engineering, and Mathematics (STEM) and other reform efforts use differences to provide evidence and support for all four STEM features. But, reform is more than information about Science, Technology, Engineering, and Mathematics. Traditional science teaching is typically defined as core disciplinary teaching that is labeled Science incorporating Technology, Engineering, and Mathematics. However, all should be considered "together" if we are to accomplish real reforms in teaching instead of merely information to be considered and used by teachers. Some even add an "A" to make this STEAM and the "power" it adds. But, these do not help with real reform for all! The focus too often remains on "what" is to be taught and remembered for testing rather than on "how" teaching should occur and provide experiences with learning.

Teachers and researchers should focus on applications of real science learnings along with mathematics as well as engineering and technology (certainly more what the human world is like for many). Engineering and technology are based on uses—and not just more information to provide for students. Too many continue teaching

Continued

information presented by workshop providers. Thus, government and organizations continue to think about "topics" and continue to develop textbooks with "things" to consider for teaching rather than focusing on "how" teaching needs to occur if students are going to experience the real "doing" of science.

STEM has certainly helped with improving school science—but it remains difficult to not focus on merely including engineering and technology as "add ons." STEM offers encouragement and understanding of science. More importantly it illustrates the actual "doing" of science, which includes engineering and technology and integrates mathematics as being needed to see the relationship of these human uses. Physicists have repeatedly pointed out that to understand physics we need to know mathematics. Technology and engineering indicate applications for "doing" science! It is not just information from books—and used in lectures. Identifying successes with actions that focus on how teachers teach as opposed to what they teach indicate specific ways teachers can improve their teaching.

The integration of S-T-E-M rather than simply the add-on of material to be covered distinguishes this educational innovation for Yager. The integration of three critical elements at the policy level distinguish STEM from failed education reform movements of the past, according to Jennifer Zinth.

Jennifer Zinth
Director of High School and STEM Programs, Education Commission of the States

State policy makers continue to adopt STEM policies and initiatives that are missing the elements necessary to ensure the quality, reach, staying power, and ultimately, return on investment of these efforts. What's missing? And how can policy makers reverse this state of affairs? State leaders would need to ensure that state-level STEM policies and initiatives include three interrelated components, namely: (1) a statewide organizing structure or statewide coordination; (2) an adequate and consistent funding stream; and (3) evaluation or quality assurance of all funded projects.

Why these three? As for a statewide organizing structure: When states adopt and implement STEM programs through an assortment of state-level entities, all too often, the left hand doesn't know what the right hand is doing, creating potential misalignment of programs or duplication of efforts within a state.

Continued

Inadequate funding may compromise program quality or fidelity of implementation, or may limit access to programs of the highest quality to a lucky few students. Without reliable funding from year to year, the best of STEM programs can lack staying power. Alternatively, an adequate and reliable funding source can support dedicated full-time employees at the designated statewide coordinating body to see to program development and implementation, and ensure the scalability of the high-quality STEM experiences we wish for all our students.

And without a process or dedicated staff to evaluate whether STEM programs are delivering the hoped-for return on investment in terms of student engagement and achievement, it's of little consequence whether programs are coordinated statewide or adequately, reliably funded.

If forced to choose a single thing about America's STEM movement that most excites me about the promise of STEM, it's arguably that everybody "gets it." In an age of polarizing (and polarized) political discourse, it's rare and refreshing to be able to identify an education policy issue like STEM that can unite policy makers across the aisle, across states. At a time where some schools complain about waning parental engagement, I hear stories of parents showing up in droves to "STEM nights" at their schools, excited to discover what their children and their peers are learning. And when exposed to high-quality STEM learning experiences, the students themselves—from the early grades through college—immediately make the connection between what they're learning and the potential they possess to improve their worlds.

Now if we could just ensure the three policy supports are there in every state to achieve the promise of STEM!

The three ingredients of Zinth's recipe for success for statewide STEM were baked into Iowa's structure out of the gate in 2011, thanks to the leadership of our "master chef," Governor Kim Reynolds. As co-chair of the Governor's STEM Advisory Council, she crafted a statewide organizing structure, a consistent funding stream, and an evaluation/feedback mechanism for quality assurance. She has scaled up the state's STEM programs, which have been documented to boost the STEM interest and performance of Iowa youth. Thus, the final word of *Creating a STEM Culture for Teaching and Learning* is reserved for Reynolds.

Kim Reynolds
Governor of Iowa

The Promise of STEM in Iowa and America

Langston Saint, a third grader at Loess Hills Computer Programming Elementary School in Sioux City, Iowa, told me that using coding for his animated social studies project makes learning more fun.

In a classroom redesigned to promote more collaborative STEM education at Hoover High School in Des Moines, teaching and learning are nothing like business as usual.

When I visited rural IKM-Manning School District on a statewide STEM tour, the power of a community supporting school–business partnerships to grow a more vibrant workforce was impressive.

The great promise of the STEM movement in Iowa—and across the nation—is its ripple effect:

STEM provides students with knowledge, confidence, and problem-solving skills that will help them succeed in their personal and professional lives. STEM provides educators with professional development that redefines the classroom, including opportunities to work with business partners so what students learn is more relevant to the real world. STEM provides employers with a more highly skilled workforce so they can innovate and expand.

As co-chair of the Iowa Governor's STEM Advisory Council, our mission is increasing student interest and achievement in STEM. That's why our focus is selecting high-quality STEM education programs that are delivered to students across Iowa from preschool through high school. We reached more than 100,000 children last school year in classrooms and other settings.

And Iowa is seeing results. Students who participated in the STEM education programs offered by the council scored higher in math, science, and reading than their peers, according to a 2016 independent evaluation. Nearly every school district in the state has had at least one of those programs.

The challenge Iowa and America must meet to fulfill the promise of STEM is ensuring all students have outstanding STEM education opportunities every year, no matter where they live. I am proud of the progress Iowa has made, but even prouder that educators, business and nonprofit leaders, parents, students, and many others recognize that we must do even better.

STEM is transforming our state and the nation. That's because STEM is not just science, technology, engineering, and math. STEM is a way of life, a way of learning, and a road to the future.

References

Beede, D., T. Julian, D. Langdon, G. McKittrick, B. Khan, and M. Doms. 2011. *Women in STEM: A gender gap to innovation.* Washington, DC: U.S. Department of Commerce, Economics and Statistics Administration. *www.esa.doc.gov/reports/women-stem-gender-gap-innovation.*

Bybee, R. W. 2013. *The case for STEM education: Challenges and opportunities.* Arlington, VA: NSTA Press.

Espinosa, L. L. 2015. Where are the women in STEM? Higher Education Today blog, American Council on Education. *https://higheredtoday.org/2015/03/03/where-are-the-women-in-stem.*

Love, T., Z. Love, and K. Love. 2016. Better Practices for Recruiting T & E Teachers. Technology and Engineering Teacher Brief, International Technology and Engineering Educators Association, Reston, VA. *www.iteea.org/File.aspx?id=95407&v=b613c163.*

National Action Council for Minorities in Engineering. 2017. Underrepresented minorities in STEM. *www.nacme.org/underrepresented-minorities.*

National Assessment Educational Progress Technology and Engineering Literacy (NAEP/TEL). 2014. Technology and engineering literacy (TEL) assessment. The Nation's Report Card. *www.nationsreportcard.gov/tel_2014.*

Neuhauser, A., and L. Cook. *U.S. News & World Report.* 2016. 2016 U.S. News/Raytheon STEM Index Shows Uptick in Hiring, Education. May 17. *www.usnews.com/news/articles/2016-05-17/the-new-stem-index-2016.*

Scheffler, I. 1960. *The language of education.* Springfield, IL: Charles C. Thomas.

U.S. Bureau of Labor Statistics. 2013. Occupational employment projections to 2022. *www.bls.gov/opub/mlr/2013/article/occupational-employment-projections-to-2022-1.htm.*

U.S. Bureau of Labor Statistics. 2017. Labor force statistics from the current population survey. *www.bls.gov/cps/cpsaat11.pdf.*

U.S. Department of Education. 2017. Perkins web portal. *https://perkins.ed.gov/pims/dataexplorer.*

U.S. Department of Labor. 2016. Data and statistics. *www.dol.gov/wb/stats/stats_data.htm.*

Epilogue

Much to the surprise of newcomers to the (trans)discipline STEM, it is hardly at all about science or technology, or engineering or math per se. If there is a singular upshot to this contribution to the field of STEM, it is that STEM assembled as an acronym sums to well more than its parts. It is a prescription for education. STEM synchronizes learning content to applications outside the school where people live and work. STEM teachers mingle education with the world of commerce. STEM engages communities in the conduct of schooling. And STEM merges the disciplines, erasing artificial boundaries born of well-intentioned efficiency but at the cost of meaning and value.

S **Synchronize** learning content to applications out there.

T **Teachers** experience the world of commerce.

E **Engage** the whole community in education.

M **Merge** disciplines to focus on challenges, not subjects.

A STEM prescription for K–16 education

Index

Page numbers printed in **boldface** type refer to figures or captioned images.

INDEX

INDEX

INDEX

NATIONAL SCIENCE TEACHERS ASSOCIATION

Creating a STEM Culture for Teaching and Learning

INDEX